An Introduction to Animal Behavior

An Integrative Approach

ALSO FROM COLD SPRING HARBOR LABORATORY PRESS

Other Titles of Interest

Evolution by Nicholas H. Barton, Derek E.G. Briggs, Jonathan A. Eisen,
   David B. Goldstein, and Nipam H. Patel

An Introduction to Nervous Systems by Ralph J. Greenspan

An Introduction to Animal Behavior

An Integrative Approach

Michael J. Ryan
University of Texas, Austin

Walter Wilczynski
Georgia State University, Atlanta

COLD SPRING HARBOR LABORATORY PRESS
Cold Spring Harbor, New York • www.cshlpress.com

An Introduction to Animal Behavior
An Integrative Approach

© 2011 by Cold Spring Harbor Laboratory Press, Cold Spring Harbor, New York
All rights reserved
Printed in the United States of America

Publisher	John Inglis
Acquisitions Editor	Alexander Gann
Development Director	Jan Argentine
Development Editor	Judy Cuddihy
Project Manager	Joan Ebert
Permissions Administrator	Carol Brown
Production Editor	Kathleen Bubbeo
Compositor	Techset Composition Ltd
Production Manager	Denise Weiss
Cover Designer	Ed Atkeson

Front cover: The integrative nature and complexity of animal behavior studies are illustrated in this example of why birds sing, in this case a singing male oropendola, a member of the New World blackbird family. Data obtained (clockwise from top, left) from studies of the morphology of the syrinx that enables complex song production, the brain circuitry involved in song learning, a phylogeny of some oropendola taxa and sonograms of the song they produce, and a hypothetical relationship between male song repertoire size and the number of offspring sired are combined to provide an answer.

Library of Congress Cataloging-in-Publication Data

Ryan, Michael J. (Michael Joseph), 1953-
  An introduction to animal behavior : an integrative approach / Michael J. Ryan and Walter Wilczynski.
        p. cm.
  Includes bibliographical references and index.
  ISBN 978-1-936113-18-7 (cloth : alk. paper) -- ISBN 978-0-879698-58-4 (pbk. : alk. paper)
  1. Animal behavior. I. Wilczynski, W. II. Title.

  QL751.R87 2011
  591.5--dc22

                        2010048432

10 9 8 7 6 5 4 3 2 1

All World Wide Web addresses are accurate to the best of our knowledge at the time of printing.

Authorization to photocopy items for internal or personal use, or the internal or personal use of specific clients, is granted by Cold Spring Harbor Laboratory Press, provided that the appropriate fee is paid directly to the Copyright Clearance Center (CCC). Write or call CCC at 222 Rosewood Drive, Danvers, MA 01923 (508-750-8400) for information about fees and regulations. Prior to photocopying items for educational classroom use, contact CCC at the above address. Additional information on CCC can be obtained at CCC Online at http://www.copyright.com/.

All Cold Spring Harbor Laboratory Press publications may be ordered directly from Cold Spring Harbor Laboratory Press, 500 Sunnyside Boulevard, Woodbury, New York 11797-2924. Phone: 1-800-843-4388 in Continental U.S. and Canada. All other locations: (516) 422-4100. FAX: (516) 422-4097. E-mail: cshpress@cshl.edu. For a complete catalog of Cold Spring Harbor Laboratory Press publications, visit our World Wide Web Site http://www.cshlpress.com/.

Copyright Disclaimer

Every effort has been made to contact the copyright holders of figures and photographs in this text. Any copyright holders that we have been unable to reach or for whom inaccurate information has been provided are invited to contact Cold Spring Harbor Laboratory Press.

*MJR dedicates this book to his daughters Emma and Lucy,
whose interest in and enthusiasm for the natural world
continues to fuel his own.*

*WW dedicates this book to Debbi Greene, who
continues to be his inspiration.*

Contents

Preface

ANIMAL BEHAVIOR WEAVES ITSELF THROUGHOUT the tapestry of biology: It is sparked when neurons fire in response to stimuli in the external world, it forms the interactions that lead to reproduction and genetic propagation, and it enhances complex group function, even when it emerges from seemingly simple self-organizing principles. Animal behavior is the most integrative endeavor of biology—its practitioners vary from those interested in the molecular, hormonal, and neural control of behavior to those who concentrate on its adaptive significance, historical patterns, and ecological consequences. As such, animal behavior provides the interface between proximate and ultimate biology—the processes internal to the animal and the consequences external to it. As an example, an animal's sensory system is exquisitely designed to perceive and organize the details of its external world, and few would argue the potency of natural selection in bringing about such an adaptation. But selection does not act directly on a sensory system; it acts on the behaviors that emerge from the interaction between the sensory system and the world around it. Nor does behavior evolve directly; rather it comes about from heritable changes in its underlying mechanism. How can we hope to understand animal behavior, or more generally, the science of life, if we commit to the parochialism of proximate versus ultimate questions and the mantra that they have nothing to do with one another? A subtext of this book is that we welcome the demise of the old proximate–ultimate dichotomy and celebrate what we hope will become a seamless segue from molecules and neurons to adaptation and evolution in the quest to understand why animals behave the way they do.

Both of our research programs have benefited from continual discussion and interactions across the proximate–ultimate boundaries of animal behavior. Walter Wilczynski was a postdoctoral associate studying the neuroscience of auditory systems in the Section of Neurobiology and Behavior at Cornell University at the same time that Michael J. Ryan was completing a thesis on sexual selection and acoustic communication in the same department. We were both hired at the University of

Texas in the early 1980s, W.W. in the Department of Psychology and M.J.R. in what was then the Zoology Department (and later morphed into the Section of Integrative Biology). We quickly began a series of research collaborations on the mechanisms and evolution of acoustic communication in anurans that continues to this day. It was through these interactions that we both became convinced that integrative animal behavior consists of more than cataloging data at different levels of analysis but is most potent when it uses information at each level of analysis to inform research and interpretations at other levels. We also became convinced that this interaction is symmetric, with proximate details being as important to understanding ultimate processes as evolution is critical for understanding the precise design of mechanisms of animal behavior. It is in this spirit that we set out to present our view of integrative animal behavior.

Our book is not intended to be an exhaustive review of the field of animal behavior but instead an introduction from an integrative perspective. It is aimed at upper-level undergraduates, graduate students, and others wanting exposure to this particular approach to animal behavior. The field of integrative animal behavior is so vast that it is impossible to cover exhaustively. Our strategy was to discuss the areas of animal behavior that both were historically of great interest to the field and continue to receive considerable interest among researchers and students. Necessarily, some interesting areas were omitted or mentioned only briefly to keep the size of the book manageable: These neglected topics include aposematic coloration, plant–animal interactions, and many areas of great interest to comparative psychologists, such as animal learning and conditioning and the growing field of animal cognition. There is a dearth of behavioral development in this book, in part because the marriage of this historically important field and the current rages in "evo-devo" and "eco-devo" seem not yet to be consummated. In those areas of animal behavior we have covered, we hope to provide the reader with a sense of their basic ideas and research problems as well as examples of both classic and contemporary research. The references we provide for each chapter are meant to do the same. We aimed to provide a collection of classic papers, new research findings, and several review papers, chapters, and books that will afford the interested reader an entry into the field as well as credit those works that provide its foundation. Unavoidably, we neglected some important and interesting work, especially from the large amount of new research coming into the field every day. We apologize in advance for any oversights.

In preparing this book, we drew heavily on others that are more specialized and provide a greater depth on aspects of behavior, neuroscience, and behavioral endocrinology. Most are included in the references in various chapters where they were most helpful, but we feel it important to express our appreciation and mention those here, as they would be excellent resources for readers who want more depth in various areas covered in our book. Every biologist who thinks herself educated should have read Darwin's *On the Origin of Species*. If interested in behavior, she should add *The Descent of Man, and Selection in Relation to Sex*, as well as *The Expression of the Emotions in Man and Animals*. The nine editions of John Alcock's *Animal Behavior, An Evolutionary Approach* have schooled many generations of behaviorists over the last 35 years, and the textbook *An Introduction to Behavioral Ecology* and the collected writings, *Behavioral Ecology, An Evolutionary Approach*, by John Krebs and Nick Davies, have performed a similar task for the subdiscipline of behavioral ecology. Few textbooks combine both the excitement of animal behavior and concise and critical analyses of theory and data as do those of Alcock and of Krebs and Davies. A new set of textbooks has recently updated the field and they fill a role similar to that of the old "Krebs and Davies." On the more mechanistic end of the animal behavior spectrum, Ralph Greenspan's *An Introduction to Nervous Systems* and Gunther Zupanc's *Behavioral Neurobiology* are complementary in providing excellent introductions to neuroscience as well as to neural mechanisms of behavior and neuroethology. Several books were especially helpful in providing background on the endocrinology of behavior: Randy Nelson's textbook *An Introduction to Behavioral Endocrinology*, Elizabeth Adkins-Regan's *Hormones and Animal Social Behavior*, the edited book *Behavioral Endocrinology*, by Jill Becker and colleagues, as well as the five-volume work *Hormones, Brain and Behavior*, edited by Donald Pfaff and colleagues. The latter is a major reference in the field of behavioral and neural endocrinology.

We also were greatly aided by numerous colleagues who provided comments and criticism of the chapters, shared unpublished data and background information, and engaged in discussions that clarified numerous issues. In particular, we would like to thank Karin Akre, Kim Hoke, and Deborah Lutterschmidt for their insightful reviews of the entire manuscript, and the Cold Spring Harbor Laboratory Press editor Judy Cuddihy for her exquisite attention to both detail and organization. In addition, we thank Elliott Albers, Greg Ball, Robert Bridges, David Cannatella, Lars Chittka, David Crews, Steve Emlen, Matthew Grober, Kim Huhman, Jonathon Losos, Robert Mason, Michael Meany, Kirsten E. Nicholson, Steve Nowicki, John Phillips, Joan Strassman, Greg Sword, David Quellar, and the late Jerry Waldvogel. They have all helped improve the quality of this work. We would also like to acknowledge the support we have been fortunate to receive for our research. Mike Ryan's work has received support from the National Science Foundation, the Smithsonian Tropical Research Institute, and the Clark Hubbs Professorship in Zoology. His department also provided a semester faculty leave to work on the book. Walt Wilczynski's research has been supported by the National Science Foundation, the National Institute of Mental Health, the Smithsonian Tropical Research Institute, and the Center for Behavioral Neuroscience, a Science and Technology Center established with funding by the National Science Foundation with continuing support from Georgia State University. We would also like to thank the past and present members of our own labs, who drove our research and our intellectual engagement with animal behavior and all the areas of science that connect to it. It is not inappropriate to thank in advance our future students and colleagues as well, for helping us explore areas of animal behavior yet to come.

M.J. Ryan
W. Wilczynski

CHAPTER 1

Questions in Animal Behavior

IMAGINE AN ANIMAL THAT DOES NOT BEHAVE. Then try to imagine being interested in an animal that does not behave. Without doubt, there are aspects of an animal's biology—its morphology, brain, and genes, its ecology and evolutionary history—that spark the interest of many scientists. But it is the animal's *behavior* that often focuses the attention of scientists and nonscientists, children and adults. One cannot imagine the success of the plethora of wildlife documentaries flooding the public airways if the animals in them did not behave.

Animal behavior is the fulcrum between the processes internal to the individual—its genetics, neurobiology, and physiology—and all that is external to it, including the environment, its social surroundings, and the other species it eats or is eaten by. Details of an animal's behavior have been shaped by millennia of selection molding them within the constraints imposed by the details of the animal's biology and endowing the animal with the capacity to respond flexibly to its surroundings in a manner that can sometimes be both exquisite and enigmatic. These factors make animal behavior one of the most integrative endeavors in biology: It encompasses both proximate and ultimate questions, how the behavior works, and why it has come to work as it does.

TINBERGEN'S FOUR QUESTIONS

In 1973 Niko Tinbergen was awarded the Nobel Prize in Physiology or Medicine, along with Karl von Frisch and Konrad Lorenz, as one of the founders of ethology, the study of animal behavior in its natural environment. In an attempt to organize this new endeavor, he codified the study of behavior in a 1963 paper, "On aims and methods of ethology," by proposing four general questions: causation, ontogeny, survival value, and evolution. Today we know these as mechanisms, acquisition,

adaptive significance, and evolutionary history. The first two questions address proximate causes and the latter two ultimate causes.

All four questions can address the same behavior. We can illustrate this by considering a male songbird perched on a treetop vigorously singing its complex song
in a feat of acoustic acrobatics. How can we explain this behavior? It depends on
what question we ask and what we mean when we ask it (Fig. 1.1).

Why does a bird produce the complex acoustic patterns we know as song? This
question can be rephrased to ask, how does he make the song? The mechanism of
singing involves accessing a neural code in the brain that drives the respiratory patterns

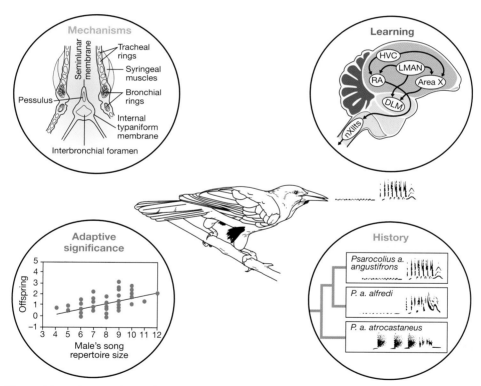

Figure 1.1. A deep understanding of animal behavior requires answers to all of Tinbergen's four
questions. One example is, "Why do birds sing?" Here we represent a singing male oropendola, a
member of the New World blackbird family, and (clockwise from *top*, *left*) the morphology of the
syrinx that enables complex song production, the brain circuitry involved in song learning, a phylogeny of some oropendola taxa and sonograms of the song they produce, and a hypothetical relationship between male song repertoire size and the number of offspring sired. (HVC) High vocal center;
(nXIIts) hypoglossal nucleus; (DLM) dorsolateral anterior thalamic nucleus; (LMAN) lateral magnocellular nucleus or the nidopallium; (RA) robust nucleus of the arcopallium.

and muscles of the bird's vocal organ, the syrinx, to produce a pattern of sound pressure fluctuations that can be interpreted by a receiver. The mechanism also involves the biomechanical details of the syrinx, which functions similarly to our larynx. There are some elegant characteristics of the songbird syrinx that endow these birds with the ability to make two separate songs at once, the "two-voiced" song, which results in some of the most complicated vocal repertoires in the animal kingdom.

We can answer this same question another way: How does he know how to make the song? We know that songbirds acquire their song through learning. To produce these complex acoustic patterns, they must be exposed to songs of another member of their species, a "conspecific," during a sensitive period early in their life. These two answers to the same question both concern proximate issues in behavior, the underlying mechanisms governing the behavior, and how the behavior is acquired.

We can also ask why the bird sings—that is, what is the function of song? Questions about the function of behavior address its adaptive significance and ask how a behavior influences the fitness of the animal producing it. Male songs have multiple functions. These songs are used to interact acoustically with other males and as courtship signals to females. Variation in song among males is under evolutionary selection because it influences a male's reproductive success. In many species, for example, females prefer males with larger song repertoires, so it seems that large repertoire size could have evolved under selection generated by female mate choice. This is a functional explanation of why birds sing complex songs, and it is rooted in understanding the potential adaptive significance of the behavior.

We can ask again: Why does the bird sing a complex song? A seemingly simple answer is: Because it is a songbird. This answer is not as glib as it might first appear; it is an informed response motivated by concerns about the past. Not all birds have complex songs. Thus, we can ask about the evolutionary history of the trait. Complex song does not occur randomly among birds. Instead, the occurrence of complex song shows a strong historical pattern. Complex song is common within one phylogenetic group of birds, the oscines or "songbirds," and is not common in other groups. This group of birds is also among the few that learn song and that have an anatomically complex syrinx. Questions about evolutionary history ask where a behavior came from. Questions of function and history are usually considered ultimate questions.

It is important to understand what questions are being asked of an animal's behavior. "Why does a bird sing complex song?" has at least four answers: because it has a complex syrinx, because it learned the song of an adult, because selection favors complex songs, and because it is a songbird. The field of animal behavior has been replete with disagreements because different researchers were asking different questions, not just because they were arriving at different answers. Tinbergen's four questions were meant to resolve some of these problems.

TINBERGEN'S ONE QUESTION

In the same 1963 paper, Tinbergen proposed a more general question that concerns all animal behaviorists: Why do animals behave the way that they do? He suggested one had to address each of his four questions to arrive at a complete understanding of animal behavior (i.e., to answer his one question). This will be a subtext running throughout this book.

Animal behavior encompasses research disciplines that sometimes have little overlap. Mechanisms of behavior are traditionally the domain of neurobiologists and physiologists, whereas studies of the acquisition of behavior are often conducted by comparative psychologists who study learning or geneticists interested in behavioral genetics. Behavioral ecologists have been at the forefront of understanding the adaptive significance of behavior, whereas studies of the history of behavior have been conducted by those with a background in phylogenetics. These are fields of science that often have little interaction, publish in different journals, and embrace different research traditions. Yet none of this makes any difference to the behavior being studied. The fact that a neurobiologist has never been in the field does not negate the importance of natural selection in shaping the neural circuit being studied; nor does the ignorance of a field biologist about the "black box" controlling behavior exclude the critical importance of this mechanism for the animal's survival.

We hope to demonstrate throughout this text that an integrative analysis of animal behavior attempts to obtain a deep and complete understanding of behavior by addressing each of Tinbergen's four questions. There are two reasons to strive for such integration. Because there are different aspects to behavior, we must address all of them to truly answer Tinbergen's one question about why animals behave the way that they do. But another important reason is that answers to questions at one level of analysis can inform answers at other levels. We can briefly illustrate this point by developing the issue of birdsong further.

One function of birdsong is to provide information about species identity. Thus, the divergence of song between populations can contribute to speciation. Although the songbird's "two-voiced" syrinx is primarily responsible for the acoustic features of the song, the morphology of the bird's beak can also influence what the song sounds like, specifically its trill rate and frequency range. But the beak not only influences what comes out of the bird but also what goes into it. Beak morphology is critical for feeding, and nowhere is this shown more clearly than in the Galápagos finches.

Many popular accounts of the voyage of the *Beagle* credit the adaptive radiation of Galápagos finches as providing Charles Darwin with insights that were critical to development of his theory of natural selection, although it did not happen until Darwin returned to England and the ornithologist John Gould drew his attention back to these birds. Although these finch species are very similar in size and color, they differ quite profoundly in the variety of their beak morphologies. Divergence in these finches was driven by diet, and their beaks evolved to complement and enable the diet. "It is very remarkable that a nearly perfect gradation of structure in this one group can be traced in the form of the beak, from one exceeding in dimensions that of the largest gros-beak, to another differing but little from that of a warbler," as Darwin stated in *The Voyage of the Beagle*. These birds evolved large, parrot-like beaks for eating fruit, grasping and probing beaks for hunting insects, and crushing beaks for crushing seeds—larger beaks for larger seeds and smaller beaks for smaller seeds (Fig. 1.2). Jeffrey Podos showed that not only does the beak's form vary with diet, but it also varies with song; males with larger beaks that eat large seeds have a slower pulse rate and more restricted frequency range than Galápagos finches with smaller beaks. Thus, selection that results in the evolution of beaks to promote feeding efficiency incidentally causes evolution of song differences among birds with different diets. It has been suggested that divergence of songs contributes to speciation between populations that not only eat different things but also sound different.

The above example uncovers a relationship between feeding ecology, song behavior, and speciation. The critical insights were how beaks influenced song and diet and how song functioned in mate recognition. We will investigate a few additional examples of such synergisms between different levels of analysis.

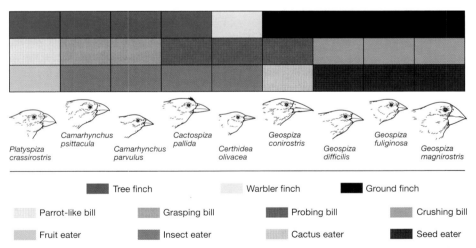

Figure 1.2. In Darwin's finches, bill morphology is closely related to diet. The *middle* of the figure shows species of all of the finches included in the analysis. Each finch (from *left* to *right*) corresponds to each column, from *left* to *right*, in the diagram above. The diagram at the *top* of the figure has three rows. The *top* row illustrates the ecological group to which each finch belongs, the *middle* row the type of bill, and across the *bottom* row a description of their diet. Each of these rows corresponds to the color-coded legend at the *bottom* of the figure. For example, the finch *Platyspiza crassirostris* is a "tree finch" (*top* row, blue) with a parrot-like bill (*middle* row, light gray) who eats fruit (*bottom* row, light blue).

SYNERGISMS BETWEEN DIFFERENT LEVELS OF ANALYSIS

Evolutionary History Informs Function

Female mate choice has been a major topic within behavioral ecology for many years. As Darwin pointed out in his theory of sexual selection, this behavior is responsible for the evolution of some of the most bizarre morphologies and behaviors in the animal kingdom. One controversial aspect of Darwin's proposal was why females should prefer certain male traits. The search for this answer usually assumes that male traits provide information to the female about the male's physical or genetic quality. However, studies of mate choice in swordtails reveal another possibility.

Mate Choice in Swordtails

Swordtails are live-bearing fishes in which the lower rays of the caudal fin are extended into a sword-like appendage. Swordtails are close relatives of platyfish, which are in the same genus (*Xiphophorus*), although platyfish usually lack these sword-like appendages. Mate choice experiments in *Xiphophorus helleri* show that females prefer males with longer swords (Fig. 1.3). Most behavioral ecology studies would stop there and speculate about what sword length tells the female about the male's health or genes. But in a series of studies by Alexandra Basolo, the focus of preference for swords shifted to a species without swords, the platyfish *Xiphophorus maculatus*. When a plastic sword was appended to a male platyfish, the females preferred the male with the artificial sword. Experiments with other platyfish and even other live-bearing fish yielded similar results. How can one interpret these results, with females preferring a male trait that does not exist in nature?

Although the phylogenetic relationships within *Xiphophorus* are not fully resolved, it appears that swordtails are all each other's closest relatives, and they inherited swords from a common ancestor after they diverged from the platyfish. However, as both swordtails and platyfish prefer males with swords, we assume that the preference was inherited from a common ancestor that would have existed before the swordtails and platyfish diverged. This historical logic suggests that the preference for the sword predated the evolution of the sword. If this interpretation is correct, it suggests that the sword was already favored by females when it evolved, as opposed to the alternative, that the sword evolved first as a signal of male quality and females then evolved a preference for it. Of course, female preferences for the trait and the traits themselves can both evolve further once the interaction of trait and preference is established.

There is more, however, to the sword's tale. In many species of live-bearing fish, females prefer larger males. It has been suggested that the sword is an energetically cheap way to make the male larger. In experiments by Gil Rosenthal and Christopher Evans using video animations, females compared two males of the same total body length (measured from the tip of the snout to the base of the tail), but one male had a sword, whereas the other male was lacking a sword. Actually, because these

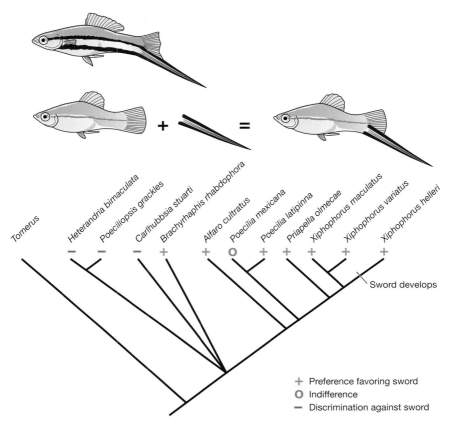

Figure 1.3. Fishes in the genus *Xiphophorus* are either swordtails or platyfish. Swordtails have elongated caudal rays that form the sword, which are lacking in platyfish. Alexandra Basolo showed that female swordtails prefer normal males with swords to males in which the sword is surgically removed, and that female platyfish prefer males with swords that are surgically added to normal swordless males. The same is true for some close relatives that lack swords. The preference for the sword therefore evolved before the sword itself. The conclusion is that males evolved traits that exploit preexisting female preferences.

were animations, the two males had the same body and the same courtship behavior. Female *X. helleri* preferred the male with the sword to the one without a sword. But when size was manipulated such that the swordless male was as long as the male with the sword, the females did not show a preference. Thus, female *X. helleri* prefer longer males whether or not a sword adds to their length. It appears that males evolved swords to exploit a preexisting female bias for larger males. These studies on the swords of swordtails provide an excellent example of how information on the evolutionary history of the behavior can inform us about how it evolved.

Mechanism Informs Function

Few natural scenes are as stunning and breathtaking as the fall foliage in northeastern North America. The trees are ablaze in riotous reds, yellows, and oranges as they prepare to drop their leaves for the winter. Plant physiologists have offered several explanations for this autumnal explosion of color. Some of the colors, such as yellow, are by-products of disintegration; as the green chlorophyll disappears, the underlying colors are revealed. But certain products (red anthocyanins) are produced only in the fall, and these products have several functions. They are antioxidants, they might function as sinks for harmful metals, and they also might warm the leaves and protect them against ultraviolet (UV) light.

Autumn Leaves and Aphid Vision

A strikingly different explanation for fall colors was offered by the evolutionary biologist W.D. Hamilton, one of the great thinkers of biology in the 20th century, and published posthumously in 2001. He posited that these are warning colors that the trees use to signal herbivores, such as aphids, to stay away. This should be an "honest" signal, that is, a reliable indicator of the signaler's quality, in that the brighter the leaves, the more toxic they are to the herbivore. The more "yellow" the leaves of a species of tree, the more likely it is to be colonized by aphids. Thus, it was suggested that these species need to invest more in warning signals to repel aphids, an argument that was independently proposed and addressed with a game theory model by Marco Archetti.

A basic assumption of the warning color argument is that the insects perceive the difference between red and yellow. The mind-bending question we ask ourselves at this point is, "When two people say they see an object they report is colored blue, how can we be sure they are really experiencing the same color as we are?" We cannot. But we can be sure that other species do not necessarily have the same experience of color that we have. An animal's sensation of color can be determined if we know the spectrum of sunlight that strikes an object, the spectrum of light reflected by that object, the spectral sensitivities of the receiver's photopigments, and the interaction of different photopigment classes in the color perception system.

Does the aphid see the same explosion of brilliance in the fall foliage that triggered such gifted poetry from Robert Frost? Hardly, according to Lars Chittka and Thomas Döring. Aphids, like all other herbivorous insects studied to date, lack a long-wavelength or "red"-sensitive cone. Their three classes of cones are sensitive to wavelengths in the very short ("ultraviolet"), short ("blue"), and medium ("green") wavelengths (Fig. 1.4). In Figure 1.4D we see how green, yellow, and red leaves from a bird-cherry tree, whose reflectances are illustrated in Figure 1.4B, would excite each of the three classes of aphid photoreceptors.

Most animals that perceive true color, or hue, do so with a color opponency system that compares the output from pairs of different cone classes. Behavioral

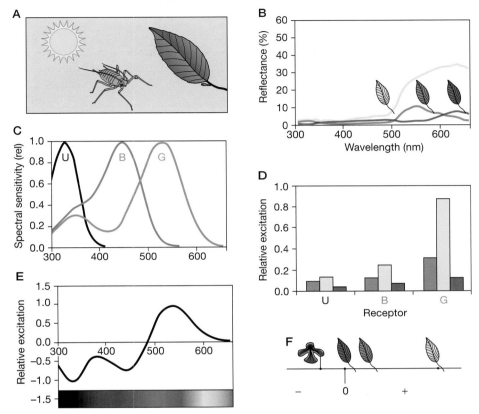

Figure 1.4. An analysis of how aphids perceive colored leaves. (A) A leaf reflects light from the sun and is seen by an aphid with green (G), blue (B), and ultraviolet (U) photoreceptors. (B) The reflectance spectra of three different colored leaves from the bird-cherry tree, *Prunus padus*. To humans, these leaves appear green, red, and yellow. (C) The wavelengths perceived by the green peach aphid is influenced by the spectral sensitivities of its UV, blue, and green photoreceptors; the tentative spectra are shown here. (D) The degree to which each of these three classes of photoreceptors are excited by the three different leaf colors varies considerably. (E) This figure shows the summed excitation of the photoreceptors. The integration of the photoreceptors is based on an opponency system with positive input from the green photoreceptor and negative input from the other two receptors. Based on behavioral data, a mechanism of this kind is presumably what many insect herbivores use as a "greenness detector." (F) The summed excitation of the color opponency system by the three bird-cherry leaves and a blue-purple flower (*Ajuga genevensis*) is illustrated on a one-dimensional scale. We see that the yellow leaf produces an even more positive signal from the "greenness detector" than the green leaf, which contradicts the behavior hypothesis that green and yellow should be perceived quite differently.

experiments suggest that aphids perceive color by an opponency system with medium versus short wavelength components, with an excitation spectrum as in Figure 1.4E. Insect herbivores generally use such a mechanism as a "greenness" detector, with positive input through the green photoreceptor and negative ones through the short wavelength receptors. Modeling this color opponency mechanism, the aphid's perception of the bird-cherry leaves and a blue flower can be represented in one-dimensional "greenness" (Fig. 1.4F). As expected, the green elicits a more positive response than red. Whereas yellow and red both appear bright to humans, red appears dull to aphids. Interestingly, insects that rely on such a color opponency system are stimulated even more strongly by yellow than by green. Hence yellow has been referred to as a "super-normal foliage-type stimulus." If this is true, then aphids should find yellow attractive and not repellent; in fact, it should be even more attractive than green. This is what was reported as data supporting the original signaling hypothesis: Tree species with more yellow had more aphids.

The suggestion that trees use bright colors as warning signals to insects was an insightful and logically consistent hypothesis about the function and adaptive significance of fall foliage. This hypothesis predicts, however, that red and yellow should be less attractive to aphids than green. But the details of the aphid's color vision suggest the opposite. The Chittka and Döring paper also notes that there are aphid species that prefer green to yellow, indicating the possibility of species differences. Nevertheless, knowledge of the mechanisms of insect color perception leads us to dismiss the hypothesis of foliage as an honest signal to herbivores.

Acquisition Informs Evolutionary History

Tinbergen's question of "evolution" asks about the contribution of evolutionary history to current behavior. In many cases, the behavioral similarities between siblings, parents, and offspring, as well as the similarities between different species, result from shared genes that influence the same behavior. The sucking and the grasping reflexes in human neonates, which are shared with all mammals and all primates, respectively, are good examples. Patterns of shared behaviors throughout a family's or a species' history often result from genomic transmission across generations.

Not all patterns of behavioral inheritance or similarity result from genomic transmission. Behaviors can be culturally transmitted across generations. Song learning in oscines is a good example. Populations of conspecific birds can have distinctively different versions of the same conspecific song—dialects. These dialects are not encoded in the genes but result from the combination of song learning and cultural drift due to different copying errors among populations.

Similarities in personality in humans, such as the extrovert–introvert continuum, can have a substantial genetic component as shown by studies of monozygotic twins. When this is the case, the evolution of the behavior in question can be analyzed using standard population genetic approaches. Whenever there are genetic contributions to behavior, however, there is sure to also be an environmental component. These gene-by-environment interactions have made the old nurture–nature debate something of an anachronism. It is even more complex when the environmental effect is personified in cultural transmission from mother to offspring.

Cross-Generational Transmission of Stress by Maternal Behavior

A remarkable study of nongenomic transmission of personality-like behavioral traits occurs in rats. There is natural variation in maternal behavior in rats including maternal licking-grooming and arched-back nursing (LG-ABN). The quantity of these maternal behaviors shown by the mother predicts the degree to which these same behaviors will be exhibited by her daughters when they later become mothers. Thus, these maternal behaviors are transmitted across generations. This transmission is not genomic, however. Cross-fostering experiments show that high-LG-ABN females reared daughters who later exhibit high LG-ABN, even though their biological mothers were low-LG-ABN females. Similarly, a pup whose biological mother was high LG-ABN but was raised by a low-LG-ABN mother later matured into a low-LG-ABN mother herself. Thus, mothering style is predicted by a pup's adopted, not its biological mother.

There is also a correlation between maternal behavior and how offspring behave. Offspring of high-LG-ABN mothers are less fearful as adults and show a more modest hypothalamic–pituitary–adrenal response in stressful situations. Cross-fostering experiments also show that the correlation between these traits is not a genetic one. Behavioral manipulation can bring about the same effect. Handling of pups increases maternal behavior toward the pups and decreases the pups' response to stress.

The influence of maternal behavior on their offspring in adulthood can also be elucidated at the genomic level. Stress reactivity is modulated by expression of genes in brain areas known to regulate the stress response. In comparison to offspring of low-LG-ABN mothers, the offspring of high-LG-ABN mothers exhibit increased serotonin expression in the hippocampus, which eventually results in the expression of a transcription factor nerve growth factor-inducible protein A (NGFI-A) (Fig. 1.5). NGFI-A then binds to a glucocorticoid receptor gene, which results in increased expression of the hormone receptor in offspring raised by high-LG-ABN mothers. Why is this not a genetic trait?

Not all inherited changes in phenotypes are due to changes in DNA sequences. "Epigenetics" refers to the class of mechanisms responsible for nongenomic inherited changes, and DNA methylation is one of these mechanisms. It plays an important role in gene expression as it can silence a gene without changing its DNA sequence. Demethylation leads to a marked expression of the gene's proteins and is a critical component of the epigenetic code. The differences in gene expression associated with stress reactivity between adult rats subjected to high versus low levels of maternal

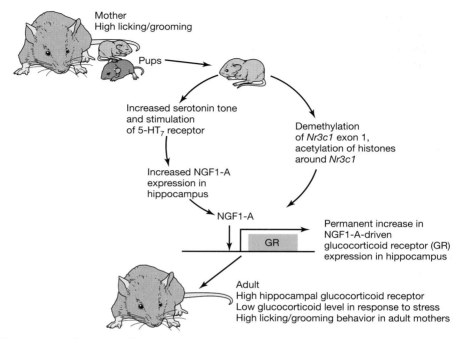

Figure 1.5. An illustration of how maternal behavior in rats is transferred epigenetically across generations. A series of traits in adult rats can be traced back to the maternal behavior that they received as pups. High maternal behavior is characterized by high levels of licking and grooming. This natal experience increases the levels of serotonin in the hippocampus, which, in turn, leads to the increased expression of the transcription factor NGF1-A. In addition, there is demethylation of the first exon of the glucocorticoid receptor (GR) gene in the hippocampus, whereas the surrounding histones are acetylated. This creates a glucocorticoid receptor gene that is permanently more open to transcriptional activation by NGF1-A, which eventually results in more glucocorticoid receptors in the hippocampus of the adult rat. It is this difference in the number of receptors that accounts for the differences between endocrine and behavioral features of adult rats that received high versus low levels of maternal behavior when they were pups.

care (LG-ABN) appear to be linked to differences in DNA methylation. These differences were shown to emerge in the first week of life, and, like the behavioral traits, they persist into adulthood and can be reversed with cross-fostering. Exposure to a high-LG-ABN mother causes demethylation of the NGF1-A binding site; however, methylation is not reversed when pups are exposed to low-LG-ABN mothers. This provides the mechanism that allows a long-term effect of NGF1-A gene expression.

These results show that maternal behavior, fear response, gene expression, and DNA methylation can all be nongenomically transmitted across generations. If the correlations between these traits within a population were examined using standard breeding value analysis based on mother–daughter correlations, or these suites of traits among species were compared using standard phylogenetic character analyses, these patterns of behavioral variation would mistakenly be characterized as having a strong genetic component, whereas in reality these patterns are evidence of cultural evolution through epigenetics and not genetic evolution.

Function Informs Mechanisms

Just as knowledge about mechanisms can inform us about function, the function of behavior can often result in the discovery of mechanisms previously unknown. Two classic examples are echolocation in bats and magnetic orientation in pigeons.

Bats Listening in the Dark

Bats have long had a place in myth and lore because of their nocturnal wanderings. This same behavior has generated curiosity from naturalists who for centuries have asked how bats can fly at such low light levels. A series of very clever experiments by Lazzaro Spallanzani and his colleagues in the 18th century in which bats were deprived of various senses demonstrated quite clearly that bats used neither vision nor odor to navigate. Spallanzani first concluded that bats must have a sixth sense that we will never understand, but later experiments by Swiss naturalist Charles Jurine suggested that they use their hearing. The famous French anatomist Georges Cuvier, however, disagreed. With apparent hubris and without the constraints of data, he declared that the patterns of air currents the bat's wings generate during flapping would be disrupted by nearby objects. The touch sensors in the bat's wings, he suggested, could then detect these disturbances and use this information for navigation. According to his theory, bats did not reckon by sight or sound, but by touch. The acceptance of this theory seemed to be based only on Cuvier's sterling reputation as an anatomist, and some of his acolytes mocked Spallanzani's hearing hypothesis with disdain, asking, "If bats see with their ears, do they hear with their eyes?"

Final resolution of this conundrum had to wait until the middle of the 20th century. Donald Griffin and Robert Galambos, novice graduate students at Harvard University at the time, made use of then recently available techniques to once again ask the question that had preoccupied Spallanzani for so long. They showed that the bat's ear was sensitive to ultrasonics (frequencies above the upper human limit of hearing, 20 kHz), that only deaf and silenced bats were unable to navigate an obstacle course, and that successful avoidance was correlated with the bat's production of echolocation signals. There seems to be little question that it was the behavior of nocturnal flight that drove scientists to finally uncover the mechanism of echolocation, which almost seems like the sixth sense that Spallanzani once attributed to bats.

Recent research by John Zook has suggested that Cuvier might have been partially correct in his assertion that bats see with their wings. Zook showed that the bats' wings are, in fact, endowed with touch sensors that contribute to navigation. Merkel cells are common mammalian touch receptors, and bat wings have Merkel-like cells with the addition of a tiny hair protruding from the center. Electrophysiological studies showed that these cells are responsive to air flowing over their surfaces. These cells can be desensitized pharmaceutically. Under these conditions, bats were capable of straight-line flight but fumbled badly when confronted with an object they needed to avoid. The philosopher Thomas Nagel asked, "What is it like to be a bat," pointing out that although we might be able to understand the mechanisms of echolocation, we could never experience it—we could never really know what it is like to be a bat. Our imaginations now are further challenged in trying to conceive what it would be like to navigate with our skin.

In the discovery of echolocation in bats, scientists were motivated by what the animals could do. In the discovery of magnetic orientation in pigeons, they were motivated by what animals could not do.

Pigeons Flying in the Fog

Pigeons are well known for their abilities to home over long distances, which is why carrier pigeons have been used to carry messages for at least 3000 years since their skill was first tapped by the Egyptians and Persians. Understanding how they do this has been the subject of many scientific careers. It was known that pigeons use the sun as an orientation cue. But pigeons can also home successfully on cloudy days. One of the preeminent researchers in this field was William Keeton of Cornell University in Ithaca, New York. Ithaca is characterized by a wealth of overcast days, which allowed Keeton ample opportunity to study homing without the sun. In a typical experiment, pigeons were driven some distance from the loft and released, and their vanishing direction at the release site was noted, as was the time that they returned to the home loft.

Although many birds use the position of the sun as an orientation cue, Keeton's pigeons usually returned to the home loft on cloudy as well as sunny days, suggesting that there must be another backup navigation system that could guide the birds homeward when the sun was not visible. There was an exception, however. If the pigeons were transported and released by one particular research assistant, the birds tended to vanish. This person drove an old Volkswagen Beetle with the motor in the back. When the cage of birds was placed in the car's back seat, the birds were adjacent to the motor. Keeton surmised that the motor's generator might be disrupting the magnetic field in the vicinity, and he decided to determine if the pigeons were using the earth's magnetic field in their orientation. He placed small bar magnets on the heads of some pigeons and non-magnetic bars on the heads of others. On sunny days the pigeons were able to return successfully but not on cloudy days. Thus Keeton concluded, and many subsequent studies have confirmed, that pigeons use magnetic cues as a backup system to their use of solar cues.

These studies of behavioral mechanisms led to a cascade of studies addressing how animals gain access to the geomagnetic field. Where is their compass and how does it work? A recent summary by Henrik Mouritsen and Thorsten Ritz points out that many birds have two sources for detecting magnetic fields. One is based on light-dependent processes in the eyes, where there are putative magnetosensory molecules, "cryptochromes." The other relies on putative magnetosensory clusters of magnetite in the upper beaks of some birds. We will return to a discussion of these mechanisms in Chapter 5.

In studies of both bat echolocation and birds, we have stellar examples of how studies of a behavior's function often inevitably lead to further questions about the behavior's underlying mechanisms.

CONCLUSIONS

Animal behavior is a phenomenon that transcends the typical categories that partition the biological sciences. To facilitate understanding between those working at different levels of analysis, Tinbergen codified the study of animal behavior with his four questions: mechanisms, acquisition, function, evolution. Tinbergen also indicated that a complete understanding of animal behavior would need to address each question. In some cases, it appears that a correct understanding of any one question depends on knowledge of the others. It is in that spirit that we offer our view of an integrative analysis of animal behavior.

BIBLIOGRAPHY

Archetti M. 2000. The origin of autumn colours by coevolution. *J Theor Biol* **205:** 625–630.

Basolo A. 1990. Female preference predates the evolution of the sword in swordtail fish. *Science* **250:** 808–810.

Chittka L, Döring TF. 2007. Are autumn foliage colors red signals to aphids? *PLoS Biol* **5:** e187. doi: 10.1371/journal.pbio.0050187.

Darwin C. 1845. *Journal of researches into the natural history and geology of the countries visited during the voyage of H.M.S. Beagle round the world, under the Command of Capt. Fitz Roy, R.N.*, 2nd ed. John Murray, London (*The Voyage of the Beagle*).

Darwin C. 1859. *On the origin of species.* Murray, London.

Galambos R. 1942. The avoidance of obstacles by flying bats: Spallanzani's ideas (1794) and later theories. *Isis* **34:** 132–140.

Grant P. 1999. *Ecology and evolution of Darwin's finches.* Princeton University Press, Princeton, NJ.

Griffin D. 1958. *Listening in the dark.* Yale University Press, New Haven, CT.

Hamilton WD, Brown SP. 2001. Autumn tree colours as a handicap signal. *Proc Biol Sci* **268:** 1489–1493.

Keeton W. 1974. The orientational and navigational basis of homing in birds. *Adv Study Behav* **5:** 47–132.

Mouritsen H, Ritz T. 2005. Magnetoreception and its use in bird navigation. *Curr Opin Neurobiol* **15:** 406–414.

Nagel T. 1974. What is it like to be a bat? *Philos Rev* **83:** 435–450.

Nottebohm F. 1996. The King Solomon Lectures in Neuroethology. A white canary on Mount Acropolis. *J Comp Physiol A* **179:** 149–156.

Podos J. 2001. Correlated evolution of morphology and vocal signal structure in Darwin's finches. *Nature* **409:** 185–188.

Rosenthal GG, Evans CS. 1998. Female preference for swords in *Xiphophorus helleri* reflects a bias for large apparent size. *Proc Natl Acad Sci* **95:** 4431–4436.

Tinbergen N. 1963. On aims and methods of Ethology. *Z Tierpsychol* **20:** 410–433.

Weaver IC, Cervoni N, Champagne FA, D'Alessio AC, Sharma S, Seckl JR, Dymov S, Szyf M, Meaney MJ. 2004. Epigenetic programming by maternal behavior. *Nat Neurosci* **7:** 847–854.

Zook JM. 2006. Somatosensory adaptations of flying mammals. In *Evolution of nervous systems* (ed. Kass JH), pp. 215–226. Academic Press, Oxford.

Function and Evolution
of Behavior

EVOLUTION IS THE CHANGE IN GENE FREQUENCIES across generations, which, in turn, can result in differences in appearance or phenotypes across the generations of a population. There are several forces that can cause evolution—mutation, genetic drift, and selection. Of these, selection is of most interest to most biologists because selection results in the evolution of functional traits. Once evolution occurs, it leaves a footprint that can sometimes be uncovered to reveal how a trait came into being. In this way, the study of evolutionary history is a natural complement to the study of selection.

This chapter provides an overview of Tinbergen's two ultimate questions—function and evolution. The function of an animal's behavior is defined by the tasks it performs. When it performs well, we assume that it enhances the animal's Darwinian fitness; that is, the behavior is favored by selection. If so, that behavior can evolve if it has a heritable genetic basis. There can also be cross-generational changes without genetic change when there is cultural transmission of behavior, as in the case of maternal behavior and stress in rats that we reviewed in Chapter 1.

Animals can evolve similar behaviors and morphologies to perform similar tasks. Flight and the wings of insects and birds are a classic example. Animals can also have similar behavior because the behaviors have been inherited from a common ancestor. A complex syrinx, complex song, and song learning in oscines are all shared through a common ancestor of the more than 4000 species of songbirds (see Fig. 1.1). But this does not mean that all songbirds sing the same song; in fact, the songs differ among virtually all songbird species. Some of these differences in song have a genetic basis, whereas other differences emerge from cultural transmission. Birds make copying errors when they learn song, and these errors can be promulgated across generations, eventually resulting in different song dialects within a species (see Chapter 7).

As noted in Chapter 1, we can ask different questions about how bird song evolved, what is the past history of the song variation, and what is the adaptive significance of these song types. As we will argue in this chapter, answers to both questions of function and evolution are needed to understand the ultimate aspects of behavior.

ADAPTIVE FUNCTION

Charles Darwin incited a revolution in science and society when he presented his theory of natural selection. In *On the Origin of Species*, Darwin argued that the history of life is a history of descent with modification, and that the modification occurs by the process of natural selection.

The process of natural selection is a simple one (Fig. 2.1). Selection results in an increase in frequency of different traits because these traits endow their bearers' with greater fitness. There are two main components of fitness—survivorship and reproduction. One must survive to reproduce, and must reproduce for genes to survive to the next generation. Selection only results in evolution if there is a transmittable component to the traits under selection; that is, for behaviors to evolve, they must be inherited. Inheritance can be genetic, if traits have a heritable basis, or cultural, if traits are transmitted by means of learning or other mechanisms. For example,

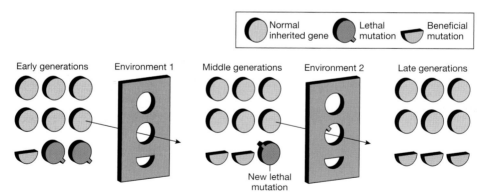

Figure 2.1. A schematic depicting the genetic changes that can take place during the process of evolution by natural selection. (*Left to middle*) Most genes exhibit allelic variation. Three classes of alleles are illustrated: the "wild-type" or normally inherited allele (gray), beneficial mutations (blue), and lethal mutations (red). Only the first two classes are passed to the next generation; the latter is screened out by the environment. (*Middle to right*) When a change in the environment favors the beneficial mutation as much as it favors the wild type, the beneficial mutation increases in frequency at the expense of the wild type, whereas new lethal mutations continue to appear at random but are weeded out of the population.

selection might favor larger size if larger animals can better defend territories and fend off predators. But size might be determined by purely environmental effects, such as food availability, with no genetic component to the variation in size; that is, size is not heritable. (Heritable is not the same as inherited. Heritability is the proportion of variance in a trait explained by additive genetic variation.) Thus, there can be selection without evolution, and in the cases of mutation or genetic drift, evolution without selection.

Figure 2.1 illustrates the joint processes of selection and evolution. Variation in alleles of genes among individuals arises by genetic mutation. Some of these variants in the population are weeded out if the bearers of those alleles do not survive and reproduce. Other alleles increase in frequency in the next generation because they endow their bearers with greater fitness. In the course of this process, lethal mutations are screened out by the environment, and beneficial ones increase in frequency. It is important to note that the environment includes the social milieu of the animal, and the animal's biotic environment can evolve as well, and thus can be a "moving target." We can refer to the aspects of the environment that generate selection as the agents of selection and the traits under selection as the targets of selection; we also refer to the trait that is the target of selection as being under direct selection.

Natural Selection and Sexual Selection, Direct Selection and Indirect Selection

Darwin's theory of natural selection was proposed to explain the evolution of functional traits, and by function he emphasized the role of survivorship. The economist Herbert Spencer first used the term "survival of the fittest" after reading Darwin, and Darwin adopted this term in his later editions of On the Origin of Species.

Survivorship is only one of the two major components of fitness; reproduction is the other. Traits that enhance survivorship will not evolve if copies of these traits are not reproduced. When Darwin reviewed variation in traits between the sexes, he was struck by a panoply of morphologies and behaviors that seemed irrelevant to survivorship. Many of them, such as the peacock's tail, seemed to hinder rather than facilitate survival. These traits shared several other characteristics: The male rather than the female usually possessed the more extreme trait variants, the traits were often more fully developed or displayed during the breeding season, and the traits were involved in breeding by enabling individuals to better defend mates or to appear more attractive to them. Darwin felt that his theory of natural selection could not explain the evolution of such traits and he developed his second great theory of sexual selection in his book The Descent of Man, and Selection in Relation to Sex. He proposed that traits that enhance an individual's ability to acquire mates would be favored by selection, even if they decrease survivorship. Such traits are thought to evolve under the countervailing forces of sexual selection and natural selection to optimize the trade-off between both forces of selection.

Some evolutionary biologists maintain Darwin's original distinction between natural and sexual selection, whereas others consider sexual selection as a type of natural selection. Both viewpoints are valid as long as it is understood that there are different components of fitness and they can be under opposing forces of selection.

Sexual selection by mate choice offers an interesting wrinkle to our understanding of trait evolution in that the agents and targets of selection can reverse roles. In many mating systems, one sex chooses the mating partner, and that choice is often based on aspects of the potential mate's phenotype. Although there are many exceptions, the canonical case is females choosing males, and in nature it is often the male that is the most adorned. In such cases, the female is the *agent* of selection, and the male trait is the *target* of selection. For example, a female's preference for males with long tails generates selection for long tails and can result in the evolution of longer tails in the population (Fig. 2.2); the female preference is the agent of selection, and male tail length is the target. The male, in turn, can then become an agent of selection and the female preference the target of selection. If males with long tails fertilize more eggs, then the female preference will also be under direct selection, and there can be the evolution of stronger preferences for longer tails.

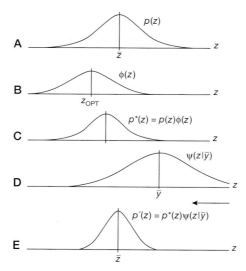

Figure 2.2. (A) The initial distribution of a male courtship trait (z), such as tail length, in the male population. (B) The probability of survival as a function of tail length. z_{OPT} is the optimal tail length for survival. (C) The distribution of the male trait after an episode of survival selection. The probability that a male survives is $\phi(z)$. (D) Females have some preference measured by y. The function $\psi(z/\bar{y})$ shows the proportional chance that a male of phenotype z will be chosen by a female. (E) The distribution of traits of the population of mated males.

When interacting individuals can be both the agents and the targets of selection, coevolution can result. Continuing cycles of coevolution, in which the evolution of one set of individuals (e.g., species, sexes) influences the evolution of another set in a continuing cycle across time, are referred to as an arms race or, more vividly, as the Red Queen effect by Leigh Van Valen, after that character's comments to Alice in Lewis Carroll's *Through the Looking-Glass* that "It takes all the running you can do, to keep in the same place." Besides the coevolution of female preferences and male traits, other examples include host–parasite interactions and the coevolution of weapons and defenses against them.

Traits can also evolve when they are not under direct selection. When a single trait is favored by selection, not only are the genes responsible for that trait passed on to the next generation but so are all the other genes in that individual. Thus traits can evolve when they are merely associated with traits favored by selection. These traits are under indirect selection. We can think of this phenomenon as genetic hitchhiking.

Female mate choice again provides an informative example. We will return to the example of preference for long tails, in which the preference is the agent of selection and the tail is the target of selection. In this example, we assume that the animals are haploid (the results are the same if diploid, just more tedious to explain), there is a single gene that determines female preference or lack of preference (P, p, respectively), and a single gene that determines male tail length (long [T] or short [t]). We also assume that males and females have genes for traits and preferences but only express the one appropriate for their sex. In a population with both forms of traits and preferences, males with long tails will be favored by selection because females with the preference mate exclusively with long-tailed males, and females without the preference mate randomly. Thus, there is an increase of T alleles versus t alleles in the next generation. There is also assortative mating as females with the preference are more likely to be mated to males with long tails than would be expected given the distribution of both preference and trait phenotypes in the population. Assortative mating also causes a bias in the assortment of alleles for traits and preferences; in the next generation P alleles are more likely to occur in individuals with T alleles. This departure of allele assortment from random expectations is called linkage disequilibrium. As female preferences select for males with long tails and cause an increase in T versus t alleles, they will also result in an increase of P versus p alleles because males with T are more likely to have P. Genes for preferences and traits are passed down to offspring of both sexes, resulting in the evolution of more males with longer tails and more females with stronger preferences for longer tails. In this scenario, which Ronald Fisher called runaway sexual selection, male traits evolve under direct selection generated by female choice, whereas female preferences evolve under indirect selection due to their genetic correlation with the male trait. We can also think of this as self-reinforcing female choice because the stronger the preference, the more evolution of both the trait and the preference.

Kin Selection, Altruism, and Selfish Genes

Evolution occurs when there is a change in gene frequencies from one generation to the next. This definition is not concerned with who bears the genes; individuals are merely vehicles for the DNA that dissolve upon death, whereas their genes are endowed with the possibility of immortality. We can couple this concept with the notion of altruism. There are many instances in animal social behavior in which individuals behave in a manner that does not increase their own direct fitness but benefits that of others. Such altruistic behaviors are most common within family groups and were often explained by notions of the good of the species. Such notions, however, usually collapse under the rigor of selection analysis.

W.D. Hamilton conceived of a theory—kin selection—that predicts that under certain conditions altruistic behavior will be favored by selection. The bottom line is whether the altruistic behavior on average promotes the transmission of copies of its genes to the next generation, with no real concern for who actually reproduces those copies. We all share many copies of our genes with other species. It is well known that we share, through a common ancestor, 98% of our genes with chimpanzees, but we also share genes through more distant ancestors: 36% with fruit flies (*Drosophila*), 15% with mustard grass (*Arabidopsis*).

This pattern of shared genetic ancestry suggests that the more genes in common, the more recent the common ancestor, and the more closely the evolutionarily relatedness between a pair of taxa. The same is true within a population. The more closely related two individuals are, the more genes they have in common. In diploid organisms, parent–offspring and sibling–sibling pairs share 50% of their genes. You would share 25% of your genes with grandparents, aunts and uncles, nieces and nephews, and 12.5% of your genes with first cousins. (Do we really share more genes with chimps than with our parents? No, these percentages refer to allelic variation of the same genes in the population, not the genes that we all have in common.) If we promote the reproduction of close relatives, we also promote the duplication of copies of our own genes; the closer the relative, the more the chance that we share genes. Hamilton's rule even predicts when altruistic behavior will be favored by selection: $c < r \times b$, where c is the cost of the altruistic act to the actor, b is the benefit of that act to the receiver, and r is the degree of relatedness between the two. The currency is fitness. A prediction, for example, is that an individual should behave altruistically toward a sibling if the act increases the sibling's fitness at least twice as much as it decreases the actor's fitness. As the degree of relatedness decreases, the greater is the discrepancy between fitness costs and benefits required for altruism to be favored by kin selection; an eightfold difference between first cousins, for example. In terms of kin selection, we predict that inclusive fitness (transmission of copies of the actor's genes by the actor as well as other individuals) rather than just individual fitness (transmission of copies of the actor's genes by the actor) will be maximized. This point is illustrated by an anecdotal quote from the geneticist J.B.S. Haldane, who is said to

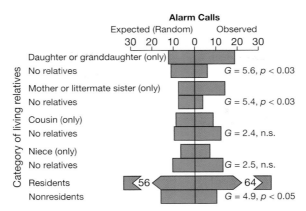

Figure 2.3. Alarm calling in ground squirrels. Summary of 102 interactions between a ground squirrel and a predator. Each paired comparison shows the number of alarm calls given by each age/sex class and the expected number of alarm calls based on how often individuals in that class were present when a predator appeared. There is a significant difference between observed and expected due mainly to females calling more and males calling less than expected.

have claimed that he would only give his life to save more than two drowning brothers or more than eight drowning cousins because he shared one-half of his alleles with each brother and one-eighth with each cousin; thus, these were the numbers needed for him to "break even" genetically.

Kin selection favoring altruistic behavior has been demonstrated numerous times, and quite elegantly by studies of ground squirrels by Paul W. Sherman. Ground squirrels are a favorite prey of many aerial predators, and squirrels give alarm calls that warn others when predators are about, even though these alarm calls increase the signalers' risk that they will become prey. Thus giving an alarm call is an altruistic behavior (Fig. 2.3). Does kin selection predict when this behavior is exhibited? As with most mammals, male ground squirrels usually disperse when they reach sexual maturity. Thus, adult females are more likely to be in the proximity of close relatives than are adult males. In this system, kin selection theory predicts that females should be more likely to give alarm calls than males, and that females should be more likely to give alarm calls when close relatives are nearby. Both predictions hold (Fig. 2.3).

Altruistic behavior also can occur in the absence of close relatedness. Reciprocal altruism occurs when an altruistic act toward a recipient increases the later benefit of an altruistic act in return. When these acts occur between species we call them mutualisms, but the same logic is involved. Bullhorn acacias lack some of the alkaloids that other plants use as chemical defenses against herbivores. These plants, however, provide "room and board" to acacia ants, which aggressively defend the tree from attacks. The ants live in hollow thorns shaped as bullhorns, and they feast on a rich

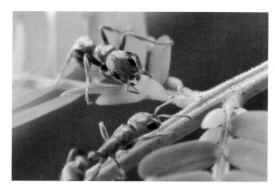

Figure 2.4. Acacia ants and Beltian bodies.

protein–lipid food that the tree provides from its Beltian bodies, which have no other function besides feeding the ants (Fig. 2.4).

The theory of kin selection refocuses selection analysis from the individual to the gene. If we follow this genic thinking to its extreme, we arrive at a notion that Richard Dawkins popularized in his book, *The Selfish Gene*. Just as through the prism of natural selection we view animals competing for resources to maximize their survival and reproduction, the prism of the selfish genes reveals intragenomic conflicts in which genes compete among themselves for representation in the next generation. A well-known example of selfish genes occurs in meiotic drive, any process in which some alleles are overrepresented in an individual's gametes. Gametes, of course, serve as the portal for genes to future generations, so an allele that is more likely to be in the gametes is also more likely to be passed on to future generations.

Meiotic drive is best studied in fruit flies in the context of segregation distortion (Sd). Sd involves two linked loci, a segregation distorter gene (*Sd*) and a responder gene. There are two forms of the responder gene, one that is insensitive (*Rsp^i*) to the products of *Sd* and one that is sensitive (*Rsp^s*; Fig. 2.5). These loci are usually found in an inverted part of the chromosome, probably because crossing-over is less likely to involve chromosomal inversion and therefore maintain the linkage between these two genes that is critical for its function. *Sd* produces a product that interferes with the normal processing of *Rsp^s* sperm; a sperm with the genotype *Sd-Rsp^s* is not functional, whereas an *Sd-Rsp^i* is able to fertilize an egg. This is why *Sd* is linked to *Rsp^i* and why these two are more likely to be found in chromosomal inversions. Selection favoring this linkage is so strong that 99% of the sperm are of the functional *Sd-Rsp^i* type. Thus, meiotic drive results in the prevalence of functional sperm that are viable for reproduction.

Sd is only one manifestation of the battle for "survival of the fittest" that takes place among genes in the same individual. It is an acknowledgment of the power of selection theory that it spans so many levels of organization.

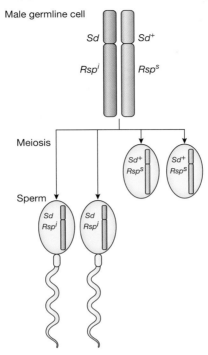

Figure 2.5. Model of segregation distorter (Sd) action in *Drosophila melanogaster*. This meiotic drive system results in the linkage of allelic variants of two genes, *Sd* and *Rsp* (responder), that promote sperm viability. There are two alleles of the *Rsp* locus that can harbor either *Rsp*^{i}, which is insensitive to distortion by *Sd*, or *Rsp*^{s}, which is sensitive to distortion by *Sd*. *Sd* interferes with the normal processing of *Rsp*^{s}-bearing sperm, although it is not clear how this happens. *Sd* and *Rsp*^{i} are often found coupled together, and thus if the homologous chromosome bears the sensitive allele *Rsp*^{s}, then sperm that bear *Rsp*^{s} degenerate, and the chromosome that bears *Sd* and *Rsp*^{i} is inherited by up to 99% of all progeny. In the diagram, the male is heterozygous, bearing *Sd* with *Rsp*^{i} on one copy of chromosome 2, and *Rsp*^{s} and no copy of *Sd* (*Sd*$^{+}$) on the other. Four meiotic products are formed, but those that bear *Rsp*^{s} fail to mature, whereas all of the functional sperm are *Sd-Rsp*^{i}.

EVOLUTIONARY HISTORY

Darwin argued that the evolution of life is a history of descent with modification and that selection explains the most interesting source of modification. Most of the similarities among species do not exist, however, because they have each evolved independently under similar forms of selection, but because they have been inherited through a common ancestor. Of course, many of these same traits did at some time in the past evolve in a common ancestor because they were favored by selection. Thus, the study of evolutionary history does not exclude the study of selection, but it offers another dimension to our understanding of why animals behave as they do.

There are at least three main contributions that studies of past history make to animal behavior: identifying patterns of change that have occurred throughout time, taking into account shared ancestry when testing hypotheses of adaptation, and estimating behaviors of species long extinct.

Evolutionary Pattern

Re-creating the history of life is not an experimental science in the sense with which most of us are familiar. Instead, this venture consists of comparing series of homologous traits among taxa and employing statistical algorithms and assumptions to propose a nexus of genealogical relationships—a phylogenetic tree. An underlying principle of phylogenetic reconstruction is parsimony, also known as Occam's (or Ockham's) razor. Parsimony dictates that with no other information available, the simplest explanation of competing explanations is most likely to be correct. Simplest refers to the number of ad hoc explanations or assumptions invoked to support a hypothesis. For example, all of the more than 5000 species of mammals have a four-chambered heart. There are at least two explanations for this fact. They all share this trait through a common ancestor, or they all independently evolved a four-chambered heart. The former is the simplest explanation, which in this case is known to be the true explanation. Parsimony is a nonparametric method that does not perform optimally under certain conditions, and many phylogeneticists rely on parametric techniques such as maximum likelihood or Bayesian inference. But the underlying philosophy is similar in that the expectation is that evolution tends to proceed parsimoniously: If two close relatives share a trait, it is likely that they do so through a common ancestor rather than that they both evolved the same trait independently. Shared ancestry thus becomes the null hypothesis that must be rejected before concluding that the similarity between the species has arisen through independent evolution. This is true even if selection is currently maintaining these traits independently in each species. This philosophy coincides nicely with the assertion of the evolutionary biologist George C. Williams, who stated that "evolutionary adaptation is a special and onerous concept that should not be used unnecessarily." The fact that characters can be shared through common descent does not exclude the role of selection; it merely brings to the forefront the other part of Darwin's theory of descent with modification. For example, no biologists would argue that the four-chambered heart is not an adaptation, only that it has not evolved independently thousands of times. The fact that the four-chambered heart is an adaptation is supported by both the physiological data demonstrating its superior function in oxygen exchange as well as the fact that the four-chambered heart has evolved independently, once in the diapsids (mammals and their extinct relatives) and once within the synapsids (birds and crocodiles, but not in lizards and snakes).

Many of the early ethologists were interested in using behaviors as characters for phylogenetic reconstruction. There was a debate as to what degree behaviors were phylogenetically informative, but that debate became moot with the widespread use of genetic sequence data in phylogenetics. If for no other reason, the abundance of DNA sequence data trumps behavior and morphology as being critical for reconstructing the history of life. Molecular phylogenies now provide the historical background to examine the patterns by which other characters have evolved.

An example using molecular phylogenies to interpret behavioral evolution is illustrated in Figure 2.6 from the work of Jonathan B. Losos and his colleagues. Dewlaps are expandable flaps of skin under the chin that many male lizards recruit into their push-up displays. In most species, dewlaps are characterized by striking patterns of color that tend to be specific to the species. The dewlap and its attendant display are best studied in lizards of the genus *Anolis*.

The phylogeny of these lizards, as illustrated in Figure 2.6B, is based on DNA sequences of seven mitochondrial genes. Superimposed on the phylogeny are classes or states of dewlap characters. The name of each species is coded to its dewlap pattern, as are branches of the phylogenetic tree. The coded branches are hypotheses of the past occurrence of each character state. Thus, we are presented with a picture of a hypothesis of the historical pattern by which diversity in dewlaps in *Anolis* lizards arose. One immediate impression, borne out by statistical analysis, is that the dewlaps are not phylogenetically informative; that is, sharing the same dewlap pattern is not indicative of close relationship. These data reject the relative importance of shared ancestry in explaining this variation in dewlap pattern among species and suggest the alternative hypothesis that selection might play an important role in how these dewlaps evolve.

Comparative Studies

The use of comparative studies to test hypotheses of adaptation has a rich tradition in ethology. Typically, the behaviors of different species with different ecologies are compared. A hypothesis of adaptation specifies how behavior and ecology should covary, and the resulting data either reject or support the hypothesis. For example, testes size, correcting for overall body size, and mating system are predicted to covary: More promiscuous males should have relatively larger testes compared to more monogamous males.

Comparative studies assume that each taxon being compared is statistically independent. In the above example, suppose there were 10 species with larger testes and promiscuous mating systems and another 10 species with small testes and monogamous mating systems. At first blush, this would seem to offer strong support for our original hypothesis. But imagine that the ancestral condition was large testes and promiscuity, there was only a single evolutionary change to small testes and monogamy, and this ancestor gave rise to 10 descendant species that shared these characters through common descent. Thus, instead of there being 20 data points in support of the hypothesis, there are only two, far too few to make any robust conclusions. We can consider this issue further examining the data in Figure 2.6B (note bracket). *Anolis bahorocoensis*, *Anolis dolichocephalis*, and *Anolis hendersoni* (the three-species clade coded in dark blue) all have dewlaps lacking pattern. They are all each other's closest relatives, so we assume that this character has been shared through descent from a common ancestor. Thus, in testing hypotheses about dewlap pattern, evolution in this clade of three species might represent only one datum.

Anolis lizards have diversified into many ecotypes. These lizards are well known for convergent evolution of morphology among species that are not closely related but inhabit similar habitats on different islands in the Caribbean, where these lizards have been studied intensely. For example, large-bodied lizards with long limbs tend to reside in the canopy, whereas smaller lizards with relatively shorter limbs tend to be

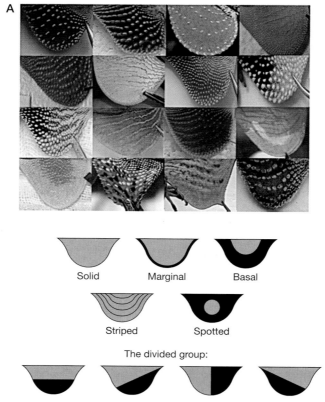

Figure 2.6. *Anolis* dewlaps. The photograph (*A, top*) shows some examples of dewlaps, and the illustrations (*A, bottom*) depict the major categories of patterns. (*B*) A phylogeny depicting the evolution of color pattern of *Anolis* dewlaps.

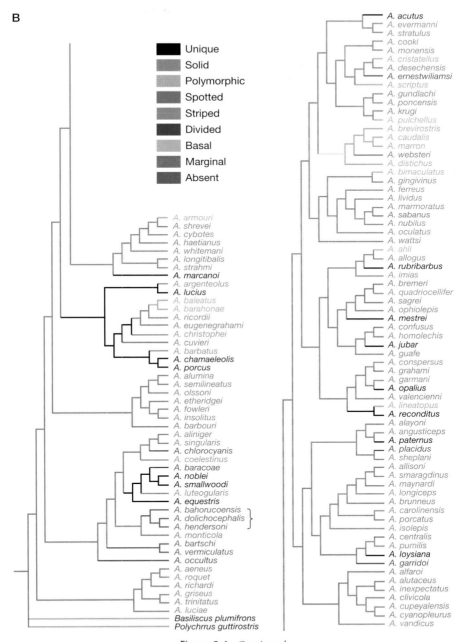

Figure 2.6. *Continued.*

on the ground. One can address a similar hypothesis about dewlaps evolving under ecological selection.

The researchers assigned each species to one of the six structural habitats in which these animals are found (grass–bush, trunk–ground, trunk, trunk–crown, crown–giant, and twig). To conduct these comparisons correctly, the variance of characters shared through common ancestry versus that thought to be due to convergent selection must be partitioned. One of the methods commonly used for this is a statistical method called phylogenetically independent contrasts. Joseph Felsenstein revolutionized the field of comparative studies of adaptation by pointing out that species are not themselves statistically independent and offered this method to partition the effect of selection and past history on differences in traits between species. This method, and other similar ones, are all based on the assumption that the more closely related are two taxa, the more likely it is that they share a similar character through common descent. In this study of *Anolis*, the researchers constructed similarity matrices for dewlap pattern, color, and size, as well as ecotype. A matrix contained all possible comparisons between pairs of species and noted simply if the two species had the same or different character. They then compared the matrices of these characters (dewlap characters, ecotype) to the matrix of phylogenetic distance, a metric of evolutionary relationships between each pair of species. The results were then used to test the hypothesis that there is a significant correspondence between dewlaps and ecology when controlling for phylogenetic relatedness. The hypothesis was not supported. We noted above that because dewlap patterns are not phylogenetically conservative traits, their divergence among taxa might be due to selection. But we can reject the hypothesis that selection generated by ecotype difference is responsible for this divergence. Why do *Anolis* dewlaps vary so much? This is a question that awaits an answer.

Estimating Ancestral Behaviors

In the book and movie *Jurassic Park*, ancient dinosaur DNA is brought to life to the joy of a paleontologist who had long speculated about the behavior of dinosaurs. Although that technology might someday be within our grasp, for now we can take an alternative approach to glimpse the behavior of taxa long extinct. We can combine data on the behavior of extant species with hypotheses of phylogenetic relationships to reconstruct ancestral behavior by using the independent contrasts method or one of the similar estimation techniques mentioned above. Toward this end, one can estimate the quantity of a trait at ancestral nodes of the phylogeny. Again the overarching principle is parsimony, and the specific goal is to minimize the proposed amount of evolutionary change across the phylogeny.

This approach was used by Michael J. Ryan and his colleagues to study the evolution of species recognition of mating calls in túngara frogs of the *Physalaemus pustulosus* species group (Fig. 2.7). Males of these frogs produce whine-like mating calls that are used as courtship signals. As with most frogs, females attend to these

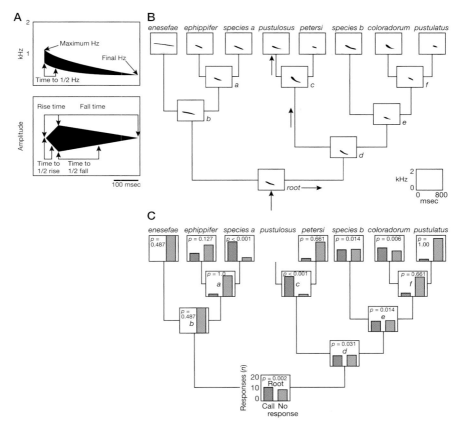

Figure 2.7. Mating calls of some species of the genus *Physalaemus*. All of these calls consist of frequencies that sweep from high to low over tens or hundreds of milliseconds. Seven call parameters are measured to characterize these calls. (A) The call parameters measured in the spectral (*top*) and temporal (*bottom*) domains. In *Physalaemus pustulosus* these parameters can be used to construct conspecific synthetic calls that females do not discriminate from natural calls, as well as synthetic calls of heterospecifics and ancestors. (B) Spectrograms of synthetic calls of eight frog species are shown on the tips of the phylogeny, and hypothesized ancestral calls are shown at all of the ancestral nodes. The black arrows show the "true" history along which one population of artificial neural networks was selected to evolve. These artificial brains were first selected to recognize the most ancestral call of the phylogeny (root). Once they were able to recognize the root call, they were then selected to recognize the call at node d. When call d was recognized, they were selected to recognize the node b call and then finally to recognize the túngara frog (*pustulosus*) call. A second set of populations was selected similarly to recognize a sequence of calls, but these were "false" histories, although the final call they were to recognize was always the túngara frog call. One such false history had the sequence: *ephippifer*, b, *coloradorum*, *pustulosus*. The various populations of artificial brains were tested with a variety of stimuli with which real female túngara frogs had been tested, and the strength of the response of the artificial neural networks and the strength of preferences of the real frogs were compared (C). The graphs represent the number of female *P. pustulosus* that responded to calls of various species and ancestors versus the number of females that did not respond. *p* tests the null hypothesis of random choice; the null hypothesis is not 0.5 but is determined empirically.

signals to identify potential conspecific mates and to discriminate against hetero-specific ones. Numerous studies of sexual communication have been conducted with *P. pustulosus*.

The mating call of these frogs can be described by measures of seven call param-eters (Fig. 2.7A). These parameters can then be used to construct conspecific syn-thetic calls that are as attractive to females as the real calls of their own males, as well as to synthesize calls of heterospecifics. Given the acoustic parameters of all the calls at the tips of the phylogenetic tree (Fig. 2.7B), ancestral calls were estimated by determining the magnitude of each of the separate call parameters for each of the ancestral nodes in a manner that minimized the total amount of change summed across the entire phylogeny (Fig. 2.7B). All of the individual parameters at each node were used to synthesize a call for that hypothesized ancestor. The heterospecific and ancestral calls were then used to test various hypotheses about the evolution of courtship signals and the receivers' recognition of them. In one case, the researchers conducted mate recognition experiments with túngara frogs: Females approach a speaker broadcasting a call if the call signals a conspecific mate. Female túngara frogs were presented a target call—a heterospecific or "ancestral" call—from one speaker and noise from another speaker. The number of recognition errors (i.e., the number of females who approached calls that were not conspecific) was quantified.

The number of recognition errors in response to each call is shown in Figure 2.7. The researchers then asked to what degree do two variables explain the number of recognition errors: acoustic similarity, the similarity between a target call and the túngara frog call; and evolutionary distance, the estimated number of DNA substi-tutions between the túngara frog and the heterospecific/ancestor of the target call. Evolutionary distance was the most important predictor, which suggested that the brain has biases in how it recognizes calls based on the history of calls that its ances-tors needed to recognize.

Steven M. Phelps tested this hypothesis by simulating brain evolution with artifi-cial neural networks. Only artificial "brains" that evolved along a history that mim-icked the history of the túngara frogs (arrows in Fig. 2.7B) could accurately predict the recognition behavior of the real female túngara frogs. These simulations showed that past history can influence the details of how current behaviors work. In this case, history does not constrain adaptations from evolving. The artificial neural net-works that had different histories still recognized túngara frog calls—they just did not do it like the real frogs did. In this case, past history biases the details of the adaptation; there are many ways to recognize a sound, and the species' ancestral legacy influences the details of how this is done. As with the study of preexisting preferences in swordtails, which we discussed in Chapter 1, this study shows that some of the understanding of why animals behave as they do might be hidden in the species' past history, in addition to our understanding derived from the behavior's current function.

CONCLUSIONS

Darwin viewed the history of life in terms of descent with modification from a common ancestor. Tinbergen's two ultimate questions address this single view. Studies of function analyze how behavior is molded by selection to maximize fitness, whereas studies of evolutionary history attempt to reveal the historical patterns by which behavior has evolved to reach its current form. These questions meld most productively when phylogenetic reconstructions are combined with data from behavioral experimentation or observation to elucidate the interaction of history and selection responsible for the behavioral diversity that so impresses many of us. Such explorations encapsulate Darwin's view of the history of life.

BIBLIOGRAPHY

Darwin C. 1859. *On the origin of species by means of natural selection*. Murray, London.

Darwin C. 1871. *The descent of man, and selection in relation to sex*. Murray, London.

Dawkins R. 1976. *The selfish gene*. Oxford University Press, Oxford.

Felsenstein J. 1985. Phylogenies and the comparative method. *Am Nat* **125:** 1–15.

Fisher RA. 1930. *The genetical theory of natural selection*. Clarendon Press, Oxford.

Hamilton WD. 1964. The genetical evolution of social behavior (I and II). *J Theor Biol* **7:** 1–52.

Maynard Smith J. 1982. *Evolution and the theory of games*. Cambridge University Press, Cambridge.

Nicholson KE, Harmon LJ, Losos JB. 2007. Evolution of *Anolis* lizard dewlap diversity. *PLoS One* **2:** e274. doi: 10.1371/journal.pone.0000274.

Orr HA. 2009. Testing natural selection. *Sci Am* **300:** 44–51.

Phelps SM, Ryan MJ. 2000. History influences signal recognition: Neural network models of túngara frogs. *Proc Biol Sci* **267:** 1633–1639.

Ryan MJ, Rand AS. 1995. Female responses to ancestral advertisement calls in túngara frogs. *Science* **269:** 390–392.

Sherman PW. 1980. The limits of ground squirrel nepotism. In *Sociobiology: Beyond nature/nurture?* (ed. Barlow GW, Silverberg J), pp. 505–544. Westview Press, Boulder, CO.

Van Valen L. 1973. A new evolutionary law. *Evol Theory* **1:** 1–30.

Williams GC. 1966. *Adaptation and natural selection*. Princeton University Press, Princeton, NJ.

CHAPTER 3

Mechanism and Acquisition of Behavior

CHAPTER 2 ADDRESSED TINBERGEN'S TWO ULTIMATE questions—evolution and function. This chapter turns to Tinbergen's two questions targeting behavior's proximate mechanisms—causation and ontogeny.

For the first of these, causation, we ask what is the "mechanism" underlying the event we call "behavior"? For even minimally complex multicellular organisms, the answers lie in the animal's nervous system. Behavior is a reflection of electrical and chemical neural processes that result in a pattern of brain activity. Other factors—hormones, genetic and genomic processes, and behavioral experience, to name three important ones—figure into a complete discussion of behavioral mechanism, but these too exert their effects by influencing nervous system operations. The functional morphology, or materials and mechanics, of parts of the body contribute to, constrain, and direct the behavioral patterns, but again the nervous system is ultimately responsible for changing static body structure into the pattern of change that equals an organism's behavior.

The second question refers to the acquisition of behavior. The original question referred to "ontogeny," or how behavior develops and changes over a life span. Behavior at any stage in an organism's life is not static. It will change as a result of life experiences, the process we call learning. But when thinking about the acquisition of behavior, an epistemological question arises: What exactly is being formed over ontogenetic development, guided by a genetic code or an epigenetic process, or acquired through adult experience? Behavior itself is ephemeral—it happens and it is over—and cannot itself be stored by the organism. Furthermore, the behavior manifests itself *after* the ontogenetic or acquisition process that prepares the organism to express it. The answer to the acquisition question directs us back to the question of mechanism. If behavior is the result of an underlying biological mechanism, then when we speak of development or acquisition, we are really speaking of an ontogenetic or experience-dependent modification of that mechanism. And, if the mechanism of behavior lies in the nervous system, neural changes must underlie the acquisition of behavior. Behavior develops or is acquired in the sense that it is the overt manifestation of some underlying change in the mechanism responsible for it.

NEURAL CONTROL OF BEHAVIOR

A nervous system exists to allow an organism to perceive the world around it and its own internal state, and to respond in an appropriate manner. It also allows the organism to monitor the results of its response and adjust itself accordingly so that any future response reflects both the world as it is and past experience (i.e., when the world and the organism were in a similar state).

Nervous systems are organized into three parts: (1) receiving systems, the sensory and perceptual components; (2) response systems, the motor systems that connect to the musculoskeletal system for overt movement and the internal organs and glands to mediate physiological responses; and (3) integrative areas that combine and compare different streams of sensory information, link sensory and motor components, and serve to adjust perception and responses in response to a host of modulatory signals reflecting internal or external circumstances, or, in the longer term, in response to stored information about past experience. In vertebrates, and in many complex invertebrates, this third component represents the vast majority of the nervous system. These three parts of a nervous system map onto the questions we ask about how an animal performs a behavior: (1) What stimuli does the animal use, and how is this information represented? (2) How are its responses generated? (3) How is the relationship between signal and response modulated over time or in reaction to current conditions?

We do not intend to present an overview of basic neural structure and function in this book; there are many excellent textbooks for that purpose. But it is important to emphasize that neural mechanisms can be considered at several biological levels. First, neurons, the cellular components of nervous systems, are electrochemical signaling and storage devices. Their electrical state is controlled by dynamic cell membrane channels that regulate the movement of ions in and out of the neuron. Neurons communicate with each other via chemical signals that pass from one to another, influencing the states of these channels, which, in turn, influence the electrical state of the receiving cell. The connection between neurons is a synapse. At a typical chemical synapse, one or at most a few types of neurotransmitter molecules are released from one neuron to bind to membrane-bound receptor molecules on the other neuron, whose actions control ion channels in that receiving neuron. In some cases, neurotransmitter action causes the interior of a receiving neuron to become more positive, and hence more likely to transmit electrical signals to other neurons; that is, it is said to have an "excitatory" effect on the receiving cell. In other cases, neurotransmitter action can cause the interior of the receiving neuron to become more negative, hence less likely to transmit electrical signals, and therefore it has an "inhibitory" effect.

A variety of amino acid derivatives serve as neurotransmitters, including glutamate (the most common excitatory neurotransmitter); gamma-aminobutyric acid (GABA) and glycine (the most common inhibitory neurotransmitters); the catecholamines dopamine, norepinephrine, and epinephrine, as well as octopamine, which is found in invertebrates; and the indolamine serotonin. Acetylcholine is another common neurotransmitter used both in the brain and at the nerve–muscle synapse. A variety of peptides and other substances are also used as neurotransmitters. In all, there are well more than 200 different molecules used as chemical signaling substances in nervous systems.

Neurons can also be electrically coupled to each other directly through specialized junctions called "electrical synapses" or "gap junctions," which allow electrical currents to pass between neurons, but these are fairly rare in vertebrates. Neurons are also sensitive to a host of other chemical signals, from hormones produced by the organism, to blood-borne chemicals such as glucose that signal physiological state, to gene products expressed within the cell itself. Therefore, one level of mechanistic analysis lies at the molecular and cellular levels and is particularly important in understanding how genetic or genomic differences impact behavior or how hormones may modulate behavioral expression, in both cases by changing the way neurons operate.

Second, neurons are interconnected through their synapses into functional circuits. These circuits define the flow of information through the nervous system. This systems-level approach helps define where in the nervous system operations take place and how information is represented on a broad spatial scale. What is remarkable about nervous systems is that cell and molecular processes are extraordinarily similar across phyla, whereas there are vast differences in their systems-level structural organization. Within a phylum—vertebrates, for example, or insects—structural similarities are quite apparent, but even with such similarities, there can be enormous quantitative differences in cell number and overall brain size. Across phyla, particularly those that have different body plans (comparing, e.g., vertebrates and coelenterates), nervous system anatomy can be qualitatively so different as to defy comparison. But in all cases, neurons act like neurons: Channel types and functions are recognizable, chemical signaling uses the same mechanisms, and, with only a few exceptions, the signaling agents (neurotransmitters, hormones) are the same.

Sensory Processing and the Perceptual World

All organisms on this planet share a common physical environment, but they can differ drastically in how they perceive it. This simple insight and its profound biological (and philosophical) significance is generally credited to biologist Jakob von Uexküll, who used the term *Umwelt* to describe the phenomenon that each organism (species;

or in its more radical form, each individual) experiences the world in its own unique way. The *Umwelt* is an organism's subjective, perceptual world. It exists and differs because every organism has a set of sensory systems that filter information from the real external world, allowing only a very restricted subset of all possible sensory experiences to be available to that organism. Nervous systems also weigh the input that is available in a manner idiosyncratic to the species. An animal's behavior is based on this restricted input and on the subjective model of the world that the animal's brain creates from it using that real-time input, its past experience, the modulations from various sources to adjust the salience of different sensory components, and the quirks and limitations inherent in neural processing. This filtered, constructed, species-specific perceptual world is what an organism knows and relies on to guide its behavior. In a very real sense, each organism exists in its own perceptual world, not in the world as it truly is.

Animal behavior studies provide numerous examples of the different perceptual worlds inhabited by diverse organisms. In some cases, the sensory systems extend our familiar experience into unfamiliar areas. Mammalian hearing is an example (Fig. 3.1). Sound, to humans, is a perception based on pressure fluctuations in the air from 20 cycles per second (or hertz, Hz) to 20,000 Hz, which we hear as low- to high-frequency tones. Elephants and other large mammals are sensitive to much lower frequencies and probably do not hear the higher frequencies we do. Their auditory perceptual world is shifted downward. Most other mammals, such as rodents and bats, hear sounds much higher in frequency than we do and are less sensitive to the low-frequency sounds we hear. Their auditory perceptual world is shifted upward. Coincident with the perceptual shift is a shift in their communication behavior, which

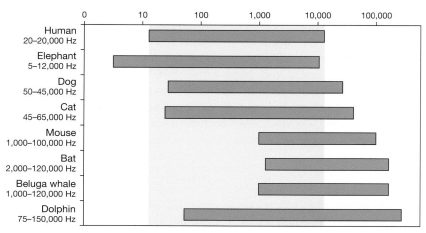

Figure 3.1. Hearing ranges of example mammalian species. The range determines the perceptual hearing world of each species.

for elephants includes the use of very low-frequency "infrasound" to communicate over long distances and for rodents the use of very high-frequency "ultrasound" for fairly short-distance social interactions such as those between mothers and pups (in both cases, in addition to the audible—to us— "sonic" sounds they make). Of course, sound is sound: Our perceptual world, our *Umwelt*, simply does not access the parts of the acoustic environment lying outside our audible range.

Most natural sound is not tonal, but instead is equivalent to a mixture of frequencies of different duration and amplitude, in different temporal, or phase, relationships to each other. Although the ear represents this complexity as its different frequency components, the brain reconstructs the individual elements into a unified percept that we "hear." In this way, we hear the natural sound—or the brain's interpretation of it—not the individual tones of which it is constructed. Ultrasound is used by bats to add a dimension to their perceptual world that we can only imagine. Bats echolocate, that is, they emit ultrasonic sound, which is reflected back to the bat. This reflected echo is detected and its features used to detect objects in the bat's environment. Bat echolocation is a far more sophisticated phenomenon than this simple description implies. Extraordinarily subtle cues in the echo are used to determine object features in addition to location extremely quickly. This is done while the bat is flying, in a way that is dynamically modified as the distance and direction from an object change. It is as though bats use their auditory system to construct a "vision" of the world around them based on sound. They inhabit this world much as we inhabit the world we construct from the ever-changing images cast on the retinas of our eyes. This is, in fact, the point: Bats inhabit a perceptual world much different from our own.

Even further from our familiar perceptual world are those animals using sensory systems that we lack. Electroreception is a sense shared by all cartilaginous and bony fish, plus a few other aquatic vertebrates. Electroreception depends on specialized sensory receptor cells located in channels or pits distributed in the skin over the head and in lines down the body. Electrosensory neurons detect electrical currents passing through their cell membranes. The passing current triggers the opening of membrane sodium ion channels to a degree proportional to the current making the cell more positive ("depolarized"), which is the signal for all neurons to actively transmit information in the form of neurotransmitter release (or, for neurons with axons, by producing action potentials for longer-range electrical transmission). In fact, *all* neurons are electosensitive to some degree in that passing a current through any neuron at a high-enough intensity will cause channel opening and ultimately neurotransmitter release. Electroreceptors are highly sensitive and are positioned to detect minute external electric currents encountered by the organism. Such currents are all around us. All animals produce electric fields because all operate using an electrically signaling nervous system that triggers movement via electrical activation of muscles. Because air is such a good insulator, the resultant electrical field extends only a negligible distance from the skin. But in a moderately good conductor like water, the electrical field can extend many centimeters beyond an organism's body. Electroreceptors in fish and sharks can thus be used to detect other organisms at a (moderate) distance based on the electrical field they uncontrollably emit. This includes potential prey, even if, for example, they are buried in the substrate or hidden from sight in murky water. This is called "passive electroreception."

Two groups of teleost fish—the African mormyrids and South American gymnotids—have taken electroreception a step further by independently evolving "active electroreception" (Fig. 3.2). Using modified muscle (or, in one case, modified neurons), these fish produce an electrical discharge (or EOD, for electric organ discharge) either in a continuous, species-specific pattern (the mormyrids) or in species-specific pulses (most gymnotids). The EODs produce an electrical field several centimeters around the fish, which the fish's own electroreceptors can detect. Such fields are distorted into local areas of higher or lower current density by objects within them, depending on their conductance. Because electroreceptors are acutely sensitive to current levels, the fish can "feel" its way through the world electrically by monitoring a current density map of its immediate surroundings. This behavioral

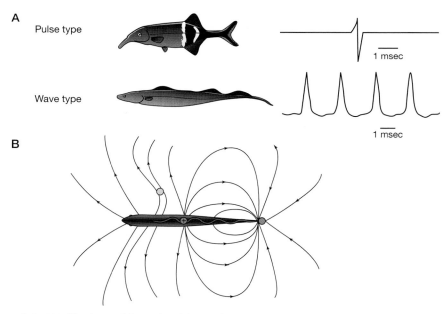

Figure 3.2. Weakly electric fish produce electric discharges from specialized muscle tissue in the tail. (A) The electric organ discharges (EODs) can be either discrete pulses as in most gymnotids (*top*) or continuous rhythmic trains or waves as in mormyrids (*bottom*). In either case, the form of the EOD is species specific. (B) The EOD produced by the electric organ in the tail causes an electric field around the fish (indicated by current lines in this schematic) that flows between the positive and negative ends of the electric organ. A conductive object in the field (blue circle) draws current toward it, thereby concentrating current density at that position. By sensing the relative strengths of the field around its body, the fish can construct a perception of the electric world around it.

modification of its perceptual world is analogous to that performed by echolocating bats and cetaceans. In each case, the animals interact with the world through the behavior of producing signals, and they then use the resultant distortion of those signals to construct a new dimension to the perceptual world. Any fish that can detect its own EOD can naturally detect the EOD of another fish. Many electroreceptive fish do, in fact, use EODs for intraspecific communication, resulting in a new dimension to this sensory system's sensory, motor, and behavioral evolution.

Electroreception is just one example of animals using sensory channels unavailable to humans to construct a perceptual world that we cannot directly appreciate, and to use that world to guide their behavior. Other examples include infrared detection in snakes; magnetic field detection in numerous migrating species; ultrasonic echolocation in bats and marine mammals; infrasound (very low-frequency sound) use in elephants; ultraviolet vision in birds, bees, and swordtails and many other fish; sensing polarization patterns in light in a variety of invertebrates and birds; and the wide range of chemical sensing by numerous and diverse taxa. All of these are cases in which sensory experiences with which we are familiar are extended or modified. Less dramatic differences in a sensory channel, either adding or subtracting components or emphasizing some parts over others, would characterize any species comparison. In all cases, the result is the species-typical *Umwelt* that von Uexküll defined, the sensory landscape in which an organism lives.

Motor Control: The Expression of Behavior

Sensory systems are the input component for neural control. Motor systems provide the output commands to the muscles, organs, and glands. Separate neurons are responsible for sensory and motor functions. For all complex organisms with a central nervous system (CNS), the CNS (in vertebrates, the brain and spinal cord) provides the interface between sensory input and motor output. Except for a few cases of the simplest reflexes where sensory neurons bringing information into the CNS connect directly with motor neurons carrying signals out of the CNS to muscles, most behavior relies on multiple neural networks within the CNS to interconnect sensory and motor neurons or influence their activity, providing the range of movements, from simple to complex, that we call behavior.

The control of movement in vertebrates can be thought of as a hierarchical series of complexity. At the simplest level are sensory-motor reflexes mediated by relatively simple circuits linking sensory input (largely from muscle and joint receptors as well as touch and pain receptors) with motor neurons whose axons leave the CNS and connect to muscles (Fig. 3.3). Such reflexes are the domain of circuits in the spinal cord (for reflex responses of the body from the neck down) or brainstem (for reflexes of face, jaw, and eye muscles). The simplest of all such reflex circuits is the monosynaptic stretch reflex, so named because it is based on a single excitatory connection in the CNS between incoming sensory neurons sensing stretch in a muscle to the motor

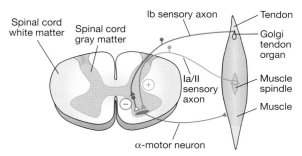

Figure 3.3. Simple reflex circuits in the mammalian spinal cord. Muscle spindles sense muscle stretch. The Ia sensory axon of stretch receptors enters the spinal cord and makes a direct excitatory connection (synapse) onto the motor neurons that innervate the muscle. When the muscle is passively stretched, the reflex triggers an automatic compensatory contraction, which helps the limb to maintain position. The Golgi tendon organ senses tension caused by the muscle contracting and pulling on the tendon. Ib sensory axons from the Golgi tendon organ enter the spinal cord and make an excitatory connection on a small interneuron. The interneuron makes an inhibitory connection on the motor neurons that innervate the muscle. When the Golgi tendon organ senses high tension, it shuts down the muscle to prevent damage.

neurons causing contraction of the same muscle. When a muscle is lengthened passively, the sensory neuron is activated proportionally to the amount of lengthening, which, in turn, activates the muscle's motor neuron proportionally, which then stimulates a compensatory contraction in the muscle. Numerous other muscle reflexes, with slightly more complex connections between sensory and motor neurons, exist within the spinal cord. Such circuits provide elements of muscle length and force control that can be accessed and used by higher-level command networks to maintain posture and position as well as build more complicated movements.

The next level of complexity is represented by the neural circuits that mediate rhythmic movements, such as those seen in locomotion and some types of sound production, as well as numerous regulatory processes from breathing to gut movements. The neural control circuits for this type of movement are often termed "central pattern generators," or CPGs. CPGs have been well described in several invertebrates and have also been studied in select vertebrate motor systems. CPG circuits are characterized as having endogenous rhythmic activity once activated. That is, they can produce a rhythmic output without depending on a rhythmic sensory input or a rhythmic input from any other neural source outside the CPG circuit itself. Their activity may also be self-sustained for a prolonged period once the circuit is activated. This intrinsic rhythm may result from the pattern of excitatory and inhibitory connections among neurons, leading to sustained, reciprocal oscillations in their activity. In some cases, a "pacemaker" neuron may exist within the circuit. Such neurons have an intrinsic rhythmic activity that is independent of any input to the neuron. They can thus act as a driver and clock for the entire circuit.

The utility of CPGs for animal behavior is obvious anytime a sustained patterned movement is advantageous. A single abrupt high-intensity input triggers the CPG underlying swimming in the gastropod mollusk *Tritonia* (Fig. 3.4). CPG activity then drives rhythmic contractions in body muscles for many seconds past the stimulus. In this way, the mechanism of the behavior (swimming) is not dependent on sustained, rhythmic sensory inputs directly driving each muscle contraction needed to move the body, and instead takes the animal away from the stimulus that triggered

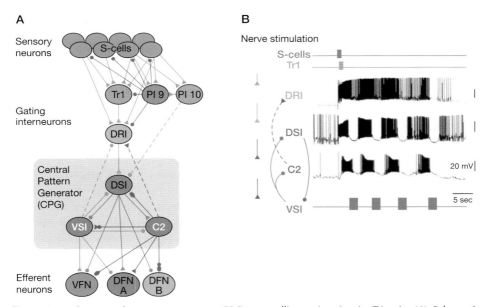

Figure 3.4. The central pattern generator (CPG) controlling swimming in *Tritonia*. (A) Schematic shows the circuit from sensory neurons (S-cells), through a variety of interneurons, to the CPG, to the motor efferent neurons. (Triangles) Excitatory synapses; (circles) inhibitory connections; (mixtures of triangles and circles) multicomponent synapses. The DSI neuron has a regular excitatory synapse on the C2, but also releases a neuromodulator, serotonin, that changes the excitability of CPG neurons. (B) A simplified version of the schematic is shown next to intracellular recordings of the neurons in the circuit. (Triangles) Excitatory synapses; (circles) inhibitory connections. Recordings show activity when levels of the neuromodulator serotonin are high, after release by the dorsal swim interneuron (DSI) neuron after strong sensory stimulation. A brief, transient activity pulse in the S-cell triggers similar brief activity in the Tr1 interneuron. The brief Tr1 activity is transformed into a longer-term steady activation in the dorsal ramp interneuron (DRI). Steady DRI activity to the DSI cell in the central pattern generator is, in turn, transformed into a rhythmic bursting pattern that is transmitted to the two other cells in the CPG. These then transmit this bursting activity to motor neurons that trigger rhythmic activity in the muscles to generate swimming. When sensory stimulation is weak, serotonin is not released, the sensitivity of C2 to DSI is reduced, and the positive feedback–induced rhythmic pattern is not triggered. Instead, the transient sensory activation is simply passed through the circuit to the efferents, resulting in a brief muscle contraction or withdrawal response without swimming away.

it—a classic example of an escape response. CPGs play a role in another behavioral domain, producing communication signals, where long-term, well-regulated patterned outputs are needed. Consider the generation of electrical signals (EODs) in the weakly electric fish. In both wave (mormyrid) and pulse (gymnotid) EOD-producing fish, the signal is highly stereotyped, very stable, and continuously produced throughout the life of the fish. EOD rhythms are controlled by a pacemaker nucleus in the midbrain that contains intrinsically rhythmic pacemaker cells. These act as a master clock for the EOD discharge connected into what is essentially a CPG network controlling the spinal cord motor neurons communicating with the electric organ. Without the intrinsic, self-generating CPG controlling the EOD, this signaling behavior would need to be driven by equally stereotyped, stable, and continuous sensory input. This is clearly impossible in the real world.

Although CPGs are intrinsically active and stable, it is important to understand that their activity can be modified by external sensory input and by neuromodulators secreted by other neurons or hormones reaching the brain. Sensory feedback, for example, is needed to adjust CPG activity and the rhythmic behavior it is controlling, to account for unexpected events or sustained imbalances or other factors that dictate compensation of the movement. For the first, consider the swimming escape response in *Tritonia*. If during the swim the mollusk brushes against a barrier on one side, it needs to adjust the swimming contractions to compensate. For the second, consider walking with only one shoe. Other relevant sensory inputs can connect with CPG circuits to start, stop, or modulate their activity and the behavior they control. The jamming avoidance response (JAR) of electric fish is another example of sensory modulation of a CPG-regulated behavior. When two wave-producing fish encounter each other, their electroreceptors will necessarily pick up a signal combining their own EOD with the EOD of the other fish. Sensing another EOD close to, but unlike, its own triggers an adjustment in the frequency of the fish's EOD discharge so that the EOD frequency moves away from the potentially interfering EOD of the other individual: This is the JAR.

The circuitry controlling the JAR in wave-producing gymnotid fish is one of the very few vertebrate behavioral-control circuits that have been described anatomically, physiologically, and neurochemically from the sensory receptors to the motor neurons and electric organ they innervate. Thus, it is useful to consider its operation in some detail. The circuit is described in Figure 3.5. The first problem the nervous system faces is detecting whether the intruder's EOD is higher or lower in frequency than the fish's own EOD signal. An electric fish senses its own EOD through skin electroreceptors that code different characteristics of the signal. There are two types of electroreceptors in gymnotid fish skin, one of which, the "tuberous receptor" has two subtypes: "P-types" that code signal amplitude, and "T-types" that code time by locking their firing to the pattern of the EOD signal. When a fish's EOD and its neighbor's combine, the P and T units see a mixed signal rather than just the fish's own EOD. The pattern of amplitude and phase modulations in the mixed

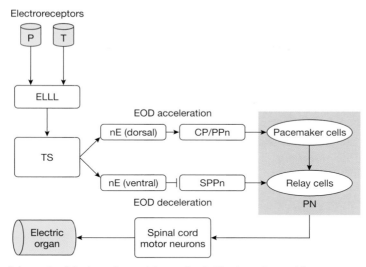

Figure 3.5. Schematic of the jamming avoidance circuit. The jamming avoidance response (JAR) is an example of a central pattern generator that is modulated by sensory input. The cells of the pacemaker nucleus (PN) are continually active, controlling the production of the species-typical electric organ discharge. Electroreceptor cells in the skin provide the sensory input, including detection of particular changes in the sensed electric organ discharge (EOD) caused by interaction with the EOD of another fish. Sensory pathways through the midbrain (TS, torus semicircularis) monitor the signals and construct specialized feature detection neurons that code whether the interference is from a signal higher or lower in frequency than the fish's own EOD. Thalamic nuclei receiving TS input serve to link sensory and motor systems and send correction signals to the PN to adjust its activity. PN cells send commands to spinal cord motor neurons that change the fish's EOD frequency to avoid interference. See text for additional explanation.

signal results in a different firing relationship between the P and T units depending on whether the neighbor's EOD is higher or lower in frequency than the fish's own EOD. Thus, the relationship between the fish's signal and the signal it wants to avoid is coded by the relationship between two separate sensory lines rather than by a single detector.

Because amplitude and phase information are detected separately, the second problem for the nervous system of gymnotid fish is synthesizing this information into a more complex percept. What one sees in the JAR circuitry is common for processing in CNS sensory systems: several sequential stages of processing moving from elemental sensory features being kept separate to gradually being combined to create neurons that code behaviorally important features of complex stimuli. In the electrosensory system, afferent fibers from the peripheral sense organs enter the brain and connect to the electrosensory lateral line lobe (ELLL) of the hindbrain, where P and T unit inputs remain separate. ELLL neurons, in turn, connect to the midbrain torus semicircularis (TS), a laminated structure in which amplitude (P) and phase (T) information is combined in several of its layers so that some neurons code the combination of amplitude and time information. These specialized TS neurons are therefore in a position to extract information about the relationship between the two EODs based on the combined patterns of amplitude (P activity) and phase (T activity). The TS sends its output to the nucleus electrosensorious (nE) in the diencephalon, where the selectivity of neurons is further refined. Here, neurons code the sign of the difference between fish and neighbor EODs: Is it above or below the fish's own EOD?

At this point, sensory analysis is sufficient to begin to address the third problem—connecting the sensory trigger to a change in motor output. For the JAR response, increases and decreases are the responsibility of different sets of cells in the nE. Sign-sensitive neurons in a dorsal part of the nE control increases in the EOD frequency. As for the sensory leg of this system, which passes through several relay points as the information ascends from lower to higher brain areas, the commands from the nE pass through several relay points as they descend from higher brain areas to the spinal cord. The dorsal nE neurons make excitatory connections on the central posterior/pre-pacemaker nucleus in the diencephalon, which, in turn, makes excitatory glutamate connections on the pacemaker cells in the pacemaker nucleus (PN), the midbrain site of the CPG controlling the EOD. When active, they gradually drive the intrinsically rhythmic pacemaker neurons to fire at a higher frequency. PN pacemaker cells connect within the PN to relay neurons. The PN relay neurons are responsible for sending the rhythmic signal to spinal cord motor neurons that directly control the electric organ in the tail. Sign-sensitive neurons in the ventral nE control decreases in EOD frequency. They connect to a different diencephalic nucleus, the sublemniscal pre-pacemaker nucleus (SPPn), and inhibit its neurons via GABA synapses. SPPn cells make excitatory glutaminergic connections on relay neurons in the PN. Inhibition of SPPn cells reduces tonic excitatory drive to the PN relay cells, which causes them to slow their output to the motor neurons.

Thus, when the electrosensory system indicates that a neighbor's EOD is lower in frequency than the fish's own, activity through the dorsal nE drives the pacemaker cells in the CPG to increase its rhythmic frequency, and the fish's EOD moves to a higher frequency, away from the lower frequency of its neighbor. When the system codes a higher neighbor frequency, activity through the ventral nE inhibits excitatory drive to the relay cells through which the pacemaker cells must communicate with the electric organ motor neurons. This then slows its responses, and the fish's EOD moves to a lower frequency, away from its neighbor's higher frequency.

Behavior beyond reflexes and the type of semiautomatic rhythmic behavior mediated by CPGs is more complex and less understood at the circuit level. Such behavior is generally goal directed, often nonstereotyped and nonrepetitive (and hence unpredictable), and, in the cases of greatest interest in animal behavior, open to rapid updating and sensory regulation. Reaching, prey strikes, homing, defensive behavior, mating, and parental behavior are all in this category. Although any of these may contain elements of reflex modification or CPG-like behavioral repetition, the complex, multidimensional nature of their expressions make them difficult to understand at a mechanistic level. At present, most of the mechanistic focus has been on understanding what sensory information guides them, what parts of the nervous system are responsible for their expression, and what factors modulate their expression.

Variation in Neural Control: Modulation and Plasticity

If behavior were driven solely by immediate sensory stimulation, most of the vertebrate brain would be unneeded. But, of course, behavior is not driven only by the stimuli an animal encounters, even in the simplest organism. Responses to identical stimuli vary on a daily and annual basis. They can differ depending on reproductive state or social context, predation risk, and the animal's energy balance. Some behaviors are incompatible: eating and fleeing a predator, for example. An animal must weigh its response options in the context of the costs and benefits of responding. This is why animals have central nervous systems rather than a simple collection of reflex loops linking sense organs directly to muscles. Large areas of the CNS are devoted to linking sensory information to motor commands through intermediate stages that allow flexibility and modification of any response that an input might trigger. Two general issues that are fundamental to the important point that behavior is not a simple stimulus–response process are modulation and plasticity.

Modulation

Neuromodulators are chemical agents that change how neurons respond to inputs, most often by changing the excitability of the neurons. Most neuromodulators are substances produced by the nervous system. The distinction between neuromodulators and neurotransmitters is a fuzzy one. Both act by binding to membrane receptors on neurons. Some substances such as serotonin can act as either neurotransmitters or as neuromodulators, depending on their mode of release and action on their targets. A main point differentiating classical neurotransmitters and neuromodulators is that neuromodulators trigger intracellular biochemical cascades in the receiving cell when they bind to their receptors, which, in turn, change the receiving cell's sensitivity to other chemical signals, its baseline level of activity, or the state of its own chemical or electrical signaling processes. In contrast to this, in traditional chemical synapses, the neurotransmitter–receptor interaction directly opens membrane ion channels, leading to an immediate but short-lived change in the receiving cell's electrical state. Neuromodulators can have widespread effects on the brain, leading to changes in overall arousal levels, or can target subregions or circuits to have effects on more specific behavioral systems. Neuromodulators include the biogenic amines— serotonin and dopamine, which are common to vertebrates and invertebrates, and epinephrine and norepinephrine, found in vertebrates, and their invertebrate analog octopamine. A variety of neurally produced peptides can also act as neuromodulators.

Many brain peptide neuromodulators are also secreted into the general circulation and serve as peptide hormones that act on other organs of the body. In some cases, the brain and body effects seem unrelated. Vasopressin, for example, is produced by neurons in the CNS of vertebrates and, as a neuromodulator released into the CNS, has profound effects on social behavior. A separate population of CNS neurons releases vasopressin into the blood, where it affects blood vessels and kidney function, controlling water retention and blood pressure. Other times a curious link is seen. Oxytocin stimulates maternal behavior when released in the brain; in the body, oxytocin released by neurosecretory neurons into the blood causes smooth muscle contraction leading to uterine contractions and milk secretion. Just as brain neuromodulator chemicals can sometimes act as bodily hormones, hormones produced by endocrine glands or other tissues can, if they are able to enter the brain, serve a brain neuromodulatory function. Numerous substances produced by the gut and fat cells, including leptin and insulin, influence hypothalamic brain areas responsible for hunger and ingestive behavior. Steroid hormones produced by the gonads and adrenal glands have effects throughout the brain. Neurons are constantly having their sensitivity levels and integrative balances adjusted by complicated mixes of brain and body-derived chemical modulators so that the relationship between sensory input and motor output is constantly varying.

Quantitative changes in behavior with varying levels of neuromodulators are easy to conceptualize. In vertebrates, increased release of dopamine in motor control regions increases movement and decreases reaction time. Although it has been a challenge to understand the exact mechanism underlying these adjustments, the neural and behavioral correlates are clear.

Work on several precisely described invertebrate circuits has revealed how quantitative shifts in neuromodulator levels can lead to qualitative shifts in behavioral outputs. This work has focused on the operation of CPGs, which, as described above, are integrated neural circuits that produce sustained rhythmic behavioral outputs, often in response to a discrete sensory input. The neural pattern emerging from a CPG is an emergent property of the interaction of many interconnected neurons. Changing the excitability of even a few of those neurons can result in a functional reconfiguration of the neural interactions so that a different neural output pattern emerges from the circuit. Swimming in *Tritonia* is controlled by a CPG (see Fig. 3.4). The swimming CPG includes two groups of dorsal swim interneurons (DSI), neurons that interconnect with other neurons in the circuit via classical chemical transmission synapses. In addition, they release a biogenic amine neuromodulator—serotonin—that acts on the neurons receiving DSI input—the C2 neurons—to increase their sensitivity to DSI input. When sensory input to the CPG is weak, serotonin is not released, the DSI is only moderately stimulatory to C2 neurons, and CPG output to the motor neurons controlling swim muscles produces a short lasting muscle contraction, seen behaviorally as a reflexive withdrawal. Stronger, sustained sensory input to the DSI neurons triggers them to release serotonin, which increases the excitability of the C2 neurons. This locks the DSI input more strongly into the CPG circuit, and the change in excitatory drive cascades through the system resulting in a self-sustained rhythmic output to the motor neurons. Behaviorally, this results in swimming away from the stimulus. In this way, increasing stimulus input shifts the functional output of the same sensory–CPG–motor neuron–muscle circuit through qualitatively different escape patterns from simple flexion of muscles (the withdrawal response) to the longer-lasting patterned flexion–extension pattern ("escape swimming").

An even more dramatic behavioral shift in bees can be traced to neuromodula-tion. As we detail in Chapter 8, eusocial insect colonies (e.g., ants, bees, wasps) have castes of sterile workers. As workers age, they move from tending to chores inside the hive to foraging outside of it. A series of elegant experiments by Gene Robinson implicates rising levels of the brain neuromodulator octopamine in this change. Foraging in older workers is triggered by brood pheromone, a chemical sub-stance derived from cuticle waxes produced by bee larvae. The antennal lobe of the brain is the major center for integrating olfactory information, including chemical signals like brood pheromone. Octopamine raises the sensitivity of antennal lobe neurons to brood pheromone and possibly other foraging-related stimuli, and, in fact, octopamine levels are high in foragers compared to in-hive workers. This type of neuromodulation is similar to the CPG modulation described above: A brain neu-romodulator changes the sensitivity of a key neuron in a circuit to a stimulus, thereby shifting the operation of a neural circuit and the behavior it controls.

A second level of modulation occurs in this circuit in that the level of brain octopamine is itself modulated by another modulator, in this case, by the bodily hor-mone called juvenile hormone (JH). When JH levels rise in the bee, octopamine levels in the brain rise correspondingly, shifting the sensitivity of antennal lobe neurons to the incoming olfactory input, thereby radically shifting the input–output function of a key olfactory-processing area of the brain and with it the behavioral repertoire of the bee. The changing cascade is part of normal development. As bees age, JH and octopamine levels rise. This developmental process can be influenced by social context. The presence of older bees retards the developmental increase in octo-pamine and the accompanying shift from nursing to foraging. How this happens is still unclear. Many other neural, hormonal, and gene expression changes coincide with the switch from nursing to foraging. Nevertheless, the JH–octopamine–antennal lobe change is a key element in the behavioral change.

Plasticity

Neuromodulation is a process whereby neural, and hence behavioral, output changes, usually temporarily, in response to current conditions. But the brain is con-figured to change in a more permanent way in response to experience. Behaviorally, we call this learning. Neurally, we call this plasticity.

Neural plasticity can take many forms. Structural changes to the nervous system can occur from the modification of existing neurons to the growth of new cells. Changes to the synaptic signaling mechanisms that link neurons into functional systems can be made, ranging from the adjustment of the sensitivity of existing connections to the formation of new ones. All such changes result from a cascade of processes that start with the activation of neural pathways and then proceed to the expression of a variety of genes by sustained or unusually intensive activation and the subsequent modification of the affected neural populations. Experiments in behavioral neuroscience using traditional laboratory animals and learning paradigms have often found these more permanent changes to depend on the synthesis of new proteins in neurons. This is consistent with the idea that many of the more permanent forms of neural plasticity involve the construction of new, or expansion of existing, synapses and an increase in the size of existing neurons.

In some instances, an increase in neuron number has been seen, pointing to the addition of new cells into existing circuits. In one of the clearest examples of this ever documented, Gloria K. Mak and Samuel Weiss showed that in a male mouse, behavioral interactions with his pups increased cell proliferation in the olfactory bulb and dentate gyrus of the hippocampus. The former brain structure is important for olfactory discrimination and the latter for long-term memory formation. Using a variety of experiments with transgenic mice, Mak and Weiss demonstrated that the increased neurogenesis induced by pup–father interactions was necessary for the ability of the fathers to recognize their offspring, an ability that extended to fathers being able to remember their offspring when they had grown to adults.

Neural plasticity is often triggered by an individual's behavior or the behavior of another organism that affects that individual. By this process, behavior influences the nervous system, which reconfigures itself to modify future behavior. In other cases, neural plasticity can be triggered by environmental or internal physiological factors. The bird song system has become a classic model system for studying the various modes and mechanisms of brain plasticity and its behavioral correlates. Anthony D. Tramontin and Eliot A. Brenowitz noted in a thorough review of plasticity in the bird song system that many areas of the forebrain control system responsible for song learning and song production change radically on a seasonal basis. Many increase dramatically in size at the beginning of the breeding season through changes in cell number, cell size, expansion of the cell dendrites that are the location of synaptic inputs on those cells, and an increase in cell spacing as those dendrites grow. The synapses themselves can grow in size. Environmental cues (such as day length), physiological factors (such as gonadal steroids), and social interactions all influence these neural changes. Here, rather than being the effect of learning to sing, the neural plasticity is thought to prepare the brain for learning and producing new songs. It is important to note that neural plasticity is a lifelong phenomenon. In the case of the song birds that learn and sing songs seasonally, the song areas wax and wane over the year, growing and shrinking with the start and termination of the breeding season. Similarly, animals are capable of learning, and unlearning, new behaviors or associations throughout life. At no point is the brain fixed.

HORMONAL CONTROL OF BEHAVIOR

Hormones, the chemical substances released by endocrine glands into the circulatory system, are important modulators of behavior via their effects on the nervous system and on the body tissues through which the nervous system expresses itself. There are two basic types of hormones: peptides, which are short amino acid chains, and steroids, which are lipid-based molecules. Many of the peptide hormones released into the circulatory system as hormones are also released into the brain by neurons and used as neuromodulators there.

Traditionally, hormone effects are dichotomized into "activational" and "organizational" effects. The first is used for those effects that are short lived and limited to the time at which the hormone is present. Activation does not imply any particular mechanism, just that the effect of the hormone is more or less time-locked. Hormones can have an activational effect by directly stimulating neurons or other target tissues—for example, the peptide hormone oxytocin activates milk secretion by causing contractions of smooth muscle in mammary ducts—or by having neuromodulatory effects on neurons and neural systems—for example, oxytocin acts in the brain to increase maternal behavior and female social bonding, probably by modulating the activity of several brain areas, including their sensitivity to social signals. Peptide hormones are known for having activational effects. Peptides bind to external membrane receptors on neurons and other cells and change the operation of those neurons rapidly but for a relatively short time.

Organizational effects refer to the more permanent structural or functional changes induced by hormones via their effect on gene expression and, subsequently, structural or physiological modification of neurons or other body tissues. Such changes long outlive the presence of the hormone. Steroids can have both activational and organizational actions. This is because steroid hormones, being lipid-soluble, can enter cells and bind to intracellular receptors (either themselves or after conversion to a metabolite within the cell). The hormone–receptor complex can then act as a transcription factor regulating gene expression in the cell, thereby having fundamental influences on cell growth and differentiation.

The dual effect of steroid hormones is illustrated by testosterone's influence on male body, brain, and behavior (a topic we return to in Chapter 6). The gonadal steroid hormone testosterone causes an increase in aggression in adult male vertebrates. This is an example of activation: Aggression is high when testosterone is high, then will drop as testosterone levels drop. Testosterone also has an organizational effect through its influence on development of the male body and brain. During male prenatal development, testosterone is high, and testosterone directs tissue development in the internal and external genitals, and through its metabolites it triggers male traits in the brain; this all leads to a male phenotype. Sometime at or around birth, gonadal hormone secretion is suppressed so that juvenile males have undetectable testosterone levels. But they remain anatomical males and remain ready to express testosterone-activated male aggression when testosterone rises again at puberty. The organizational changes induced by the hormone during prenatal development persist past the presence of the testosterone that caused them, and for primary and many secondary sex characteristics are truly permanent.

The dual organizational–activational function of testosterone also illustrates the way steroid hormones orchestrate complex behavioral functions at multiple levels. Consider male reproductive behavior. Gonadal steroids control the growth of primary sex characteristics, the internal and external genitals that make reproduction possible; control the growth of secondary sex characteristics, the body structures that are used by females to identify and choose among males; and activate courtship and mating behavior that makes use of both body structures to attract females and inseminate them. The line between activation and organization can be blurry, but considering hormone effects in these two categories is useful as a heuristic when trying to understand mechanisms.

Although a substantial body of research emphasizes hormone effects on the brain, it is important to note that endocrine release, including the release of sex steroid hormones like testosterone, is controlled by the brain. In vertebrates, the hypothalamus regulates hormone levels via the release of peptides called releasing factors or releasing hormones into veins feeding the pituitary gland, which, in turn, releases peptides called trophic hormones into the general circulation. Trophic hormones trigger endocrine cells to release their peptides or steroids. The brain is thus controlling its own hormonal activation and organization, as well as its own development and expression of behavior.

GENETIC CONTROL OF BEHAVIOR

Understanding the genetic basis of behavior is important for answering questions about both the mechanism and the acquisition of behavior. This understanding is also important for addressing Tinbergen's fourth question about behavior—evolution—that was reviewed in the previous chapter. The basis for an argument about the evolution of behavior is that behavior is heritable (i.e., it has a genetic basis).

If we are dealing with the genetic basis of behavior, there are three levels where variation can contribute to behavioral differences: genomes, gene variants, and gene expression. Different genes may exist in different species or individuals, leading to differences in behavior. The most relevant cases relate to differences in taste, smell, or other sensory genes, which translate to behavioral taste preferences. This has been investigated in *Drosophila* species, which have long been models for investigating the genetic basis of behavior. Like many insects, *Drosophila* use chemical cues to detect and discriminate potential food sources. Two sister species, *Drosophila simulans* and *Drosophila sechellia*, are, respectively, generalist and extreme specialist foragers. *D. sechellia* feeds on fruit of the shrub *Morinda citrifolia* in its native Seychelles islands. It has a strong preference for the chemical toxins in this plant that are strongly repellant to other flies. Carolyn McBride characterized the genes within the olfactory and gustatory receptor families that code for the chemoreceptors underlying food sensing in *Drosophila* and compared the relative rates of gene loss and amino acid replacement mutations in the two species. The genes of these two families are expressed very narrowly in the antennal chemosensory cells that *Drosophila* use for processing chemical stimuli. The specialist species showed 10 times the rate of gene loss and had a much higher rate of replacement mutations in remaining chemoreceptor genes (compared to silent mutations). Presumably, the result is that the genome of the specialist flies contains many fewer chemical-sensing receptors and those that do exist have changed to a specialized function, resulting in an olfactory system keyed to a smaller and very different range of odors guiding food detection and identification. This remains speculative because the true functions of the myriad genes are largely unknown. Nevertheless, even though the genetic and behavioral changes are correlative in nature, they do suggest that changes in the genome could lead to very different behavioral profiles.

Gene variants (different alleles of shared genes) can map onto behavioral differences as well. Again, studies of *Drosophila* provide an example. The *period* gene is part of the molecular control system for circadian rhythms in *Drosophila* and a host of other organisms. In addition, the *period* gene controls a different type of rhythm—the rapid, rhythmic species-specific courtship song controlled by a CPG in the fly's nervous system. Species variation in a small coding region of the *period* gene's longest exon, differences of perhaps as little as four to five amino acid substitutions, is responsible for the species differences in courtship song frequency. In this case, the gene–behavior relationship can be confirmed experimentally. Mutations of the *period* gene change the pattern of the male's pulsed call (Fig. 3.6). Furthermore, *D. melanogaster* males carrying a chimeric construct with the *D. simulans* Thr-Gly region of the *period* gene produce courtship songs with rhythms more like *simulans*.

Last, differences in gene expression can correlate with behavioral differences, particularly when considering within-species differences, or even differences within an individual over time. Gene expression differences can emerge because of variants in other genes, but can also result from modulation by transcription factors by any number of extrinsic influences. For example, testosterone acts as a transcription factor in male mammals, modulating the expression of a variety of structural and functional genes that are identical in males and females. But because males have more testosterone, gene expression differs in males and females, leading to different phenotypes. Gene expression patterns can change as part of a developmental process, something

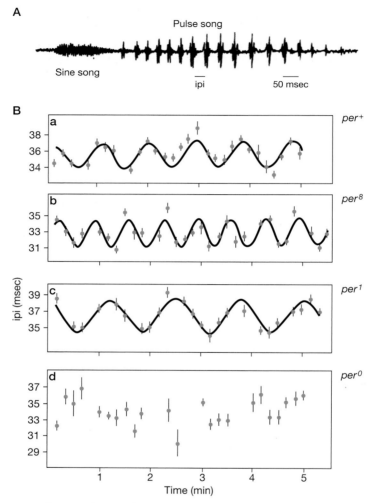

Figure 3.6. (*A*) Oscillogram of a courtship call of *Drosophila melanogaster* males. *Drosophila* species calls vary in interpulse interval (ipi, the time between pulses in the call) and rhythmic modulation of ipi. (*B*) Courtship song ipi and modulation profiles of various *per* mutants and a *per*+ male (yellow on a Canton-S genetic background). (a) *per*+, $p = 56$ sec; (b) *per*8, $p = 40$ sec; (c) *per*1, $p = 76$ sec; (d) *per*0, a *per* knockout shows a disrupted pattern.

that happens in bees that go through different life stages from in-hive nurse to external forager.

It is important to note that changes in gene expression are also connected to the neural, and consequently behavioral, plasticity discussed above. If persistent structural changes are to occur in cells, a host of genes must be up- or down-regulated to direct the molecular construction needed for these changes. In fact, activating (depolarizing) neurons for long periods induces genomic responses. A class of genes called "immediate early genes" is rapidly expressed whenever a neuron is depolarized. IEGs code transcription factors, which control the expression of other genes in the cell. It is believed that some are up-regulated for cellular housekeeping associated with the metabolic demands of increased neuron activity, but others may ultimately result in fundamental changes in neurons that underlie long-term neural plasticity of the kind related to learning. It is important to understand that such differences in gene expression are conceptually quite different from differences in genetic makeup in that they are not heritable. For example, expression patterns resulting from species differences in a regulatory gene or portion of the expressed gene can be inherited, and thus subject to selection. Individual variation in expression due to extrinsic factors such as steroid hormone levels or experience may have profound effects on an animal's behavior, but is not in itself heritable. As we discussed in Chapter 1, however, epigenetic phenomena, such as cross-generational transmission of maternal and stress-related behaviors, can occur in the absence of heritable differences in behavior.

EXPERIENTIAL AND ENVIRONMENTAL CONTROL OF BEHAVIOR

An animal's interaction with its environment—including its behavior and its interactions with other behaving individuals—is an important determinate of its future behavior. This is the essence of learning, and like the topic of genetic correlates of behavior, represents a major component of both the mechanism and acquisition of behavior that formed Tinbergen's two proximate questions. The tension between "learned" and "innate" animal behavior defined the decades-long separation between the American tradition of comparative psychology and the European conceptualization of ethology, as well as the now deservedly clichéd "nature versus nurture" debate. But no animal with a nervous system is unaffected by its genetic endowment or is incapable of learning from experience. And many complex natural behaviors are influenced by experience in surprising ways.

One of the unexpected behavioral influences on what would seem to be fixed behavioral responses is seen in the phenomenon of mate-choice copying, a phenomenon in which females tend to prefer to mate with a male that they had previously observed mating with another female. The advertisement and courtship signaling between males and females, and the consequential behavioral responses, are important components of mate choice and reproduction, as we will discuss in Chapter 7. These signals are species-specific and, as successful reproduction depends on these species-specific traits, they are under strong selection. Many of the signaling behaviors appear to be, in the terminology of classic ethology, sign stimuli, and the responses fixed action patterns, and thus have often been treated as fixed traits of an individual. In 1992, however, Lee A. Dugatkin showed that female mate choice in guppies was influenced by the choices she observed other females make. This mate-choice copying phenomenon can be strong enough to reverse a female's previously demon-strated preference for a male. It has now been demonstrated in several different taxa and in at least one case, with sailfin mollies, mate-choice copying is shown to occur in the wild and thus is not merely a laboratory artifact (Fig. 3.7).

Certainly there are heritable (genetic) components to mate signaling and responses, as the *Drosophila* example in the previous section showed and as will be discussed in Chapter 7. What the mate-choice copying literature shows is that a behavior as fundamental and evolutionarily important as mate choice can also be influenced by social, observational learning. Here, a behavioral mechanism, learning by observing the behavior of other individuals, is necessary to complete an under-standing of the mechanism and the acquisition of behavior. This understanding also has implications for the evolution of mate choice, even though the learned compo-nent of the preference is not heritable. Rather, the behavioral mechanism contributes to the variance in mating success enjoyed by both males and females. Theoretical

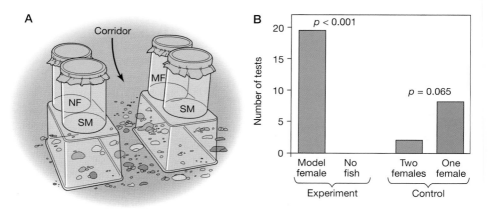

Figure 3.7. (A) Experimental setup for mate-copying experiment. Clear jars are placed under water on either side of a corridor through which a fish can swim. Jars have net tops and are filled with river water. Stimulus males (SM) are placed in two jars, the model female (MF) is in one jar next to one of the stimulus males, and the fourth jar next to the other stimulus male has no fish (NF). Female pref-erences were assessed as females swam through the corridor from the side with the empty jar and the model male (indicated by an arrow); a "preference" was indicated by the female stopping within one body length of a male. (B) Graph of female mate-choice test results. On the *left* of the graph, the y-axis indicates the number of tests (out of 20) in which more females associated with the stimulus male next to the model female or with the stimulus male next to the jar with no fish. Results from the control condition in which females rather than males were placed in the jars are on the *right* of the graph; bars indicate the number of tests in which more females associated with a stimulus female next to another female (green bar) or with a stimulus female next to an empty jar (blue bar). The graph shows that females prefer to associate with males they had seen with another female, but had no strong preference for females they had seen with another female.

models that incorporate non-independence of female decisions show that even weak mate copying dramatically and significantly affects sexual selection on male traits. By enhancing the success of successful males and decreasing the success of others, female mate copying greatly increases selection on males.

CONCLUSIONS

Questions about mechanism and acquisition can be addressed at many levels. We can demonstrate that behavior is acquired through genetic or behavioral means, and that this acquisition, particularly during ontogeny, may involve the influence of hormones. When we examine mechanisms, what is "causing" the behavior to occur, we can say that there is some hormonal, genetic, experiential, or environmental factor that triggers, directs, or modulates the behavior's appearance. In many sections of this book, we will treat proximate questions at this level. But if we drill deeper into the meaning of statements such as "a behavioral trait is inherited," or "increasing hormones increase behavioral expression," we need to realize that something is missing. Genes do not really code "behavior," they encode protein structure. Hormones do not actually increase "behavior," they bind to receptors and change the operation of cells. The behavior of one animal is not directly and causally connected to the behavior of another per se. In every case, something must intervene so that the proteins coded by genes, the receptors activated by hormones, or the behavior expressed by another organism is connected to the behavior expressed by the animal. As noted at the beginning of this chapter, that intervening structure is the nervous system, the body organ that links the external and internal environment to a behavioral response through the activity of its constituent cells and their connections to the muscles of the body.

BIBLIOGRAPHY

Dugatkin LA, Godin J-GJ. 1992. Reversal of female mate choice by copying in the guppy (*Poecilia reticulata*). *Proc Royal Soc B* **249**: 179–184.

Katz PS, Frost WN. 1996. Intrinsic neuromodulation: Altering neuronal circuits from within. *Trends Neurosci* **19**: 54–61.

Kyriacou CP, Hall JC. 1980. Circadian rhythm mutations in *Drosophila melanogaster* affect short-term fluctuations in the male's courtship song. *Proc Natl Acad Sci* **77**: 6729–6733.

Mak GK, Weiss S. 2010. Paternal recognition of adult offspring mediated by newly generated CNS neurons. *Nat Neurosci* **13**: 753–758.

McBride CS. 2007. Rapid evolution of smell and taste receptor genes during host specialization in *Drosophila sechellia*. *Proc Natl Acad Sci* **104**: 4996–5001.

Metzner W. 1999. Neural circuitry for communication and jamming avoidance in gymnotiform fish. *J Exp Biol* **202**: 1365–1375.

Schultz DJ, Barron AB, Robinson GE. 2002. A role for octopamine in honey bee division of labor. *Brain Behav Evol* **60:** 350–359.

Tramontin AD, Brenowitz EA. 2000. Seasonal plasticity in the adult brain. *Trends Neurosci* **23:** 251–258.

Witte K, Ryan MJ. 2002. Mate choice copying in the sailfin molly, *Poecilia latipinna*, in the wild. *Anim Behav* **63:** 943–949.

Chapter 4

Foraging
Sensing, Finding, and Deciding

A LL LIFE ON OUR PLANET IS DEPENDENT ON A STAR about 150 million kilometers away. Life on Earth is fueled indirectly from energy produced by our sun and is initially stored in plants, where it becomes available to some animals in the form of food. As animals eat plants and these animals are eaten by other animals, the sun's energy is passed through food webs to eventually nourish every living thing on this planet. How animals sense their food and find it, the decisions they make about whether to eat it, and the influences all of these decisions have on their fitness are all part of the study of animal foraging.

SENSING AND FINDING FOOD

Whether a grazing herbivore, a predatory carnivore, a generalist omnivore, or a specialist parasite, there are common elements to an animal's behavior—finding food, capturing it, and ingesting it. There are also common decisions—whether or not to eat, whether to eat now or save the food for later, and upon eating it, making decisions about the food's palatability and profitability. There is also a common problem—prey (even plants to herbivores) are unwilling participants in this behavior and have evolved mechanisms to thwart the predators.

Appetitive versus Consummatory Behavior

In the early 20th century, both ethologists and neurobiologists struggled with how to classify behavior and to relate stereotyped reflexes or other fixed outputs, which ethologists called "fixed action patterns," to the more common and variable goal-directed behavior that is unpredictable, nonrepetitive, and more open-ended. In 1917 Wallace Craig distinguished between "appetitive" and "consummatory" behaviors (similar in

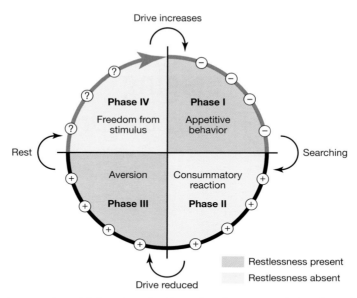

Figure 4.1. Conceptual model, based on the ideas of Wallace Craig, of the relationship of drives, appetitive behavior, consummatory behavior, and resolution during natural behavior.

concept to what Charles Scott Sherrington proposed in 1906, using different terms, in considering the neurobiology of movement) and articulated how they are related in natural animal behavior (Fig. 4.1). Appetitive behaviors are those that bring an organism toward satisfying a need or drive; they are relatively unconstrained and variable in their form. Consummatory behaviors are those that satisfy the need and reduce the motivation for appetitive behaviors; they are more stereotyped and fixed in duration. Although the words imply a connection to eating, Craig's original ideas applied to all behavior. Appetitive behaviors are searching activities, and foraging is a classic example. It is triggered by a drive or need in the organism (the need for nutrients, as signaled by internal physiological changes) and it is open-ended, continuing until the drive is reduced (by eating). Behaviorally, it has a variable, trial-and-error quality to it. Prey capture and eating are consummatory behaviors. They are far more stereotyped, stimulus-triggered behaviors that ultimately reduce the drive that caused the animal to search for food (forage).

FORAGING AND PHYSIOLOGICAL DRIVE: THE APPETITIVE PHASE

Foraging is a classic appetitive behavior in that it is driven by an internal motivation or drive—hunger. Surprisingly, there is still uncertainty about what triggers hunger. There are, however, several physiological candidates and neural signaling systems that have been implicated. Fasting triggers the release of numerous endocrine and metabolic products into the blood, and several brain neuromodulator systems, most notably neuropeptide Y (NPY), are up-regulated simultaneously. Many interacting substances likely regulate hunger and the foraging behaviors that follow it.

A clever set of experiments by Timothy Bartness and colleagues investigated the role of one peptide hormone, ghrelin, in fasting-induced foraging. Ghrelin is released by endocrine cells in the stomach, and its circulating levels increase before eating and after fasting. Furthermore, when given exogenous ghrelin, humans sense hunger. Bartness developed an experimental procedure to quantify appetitive behavior in Siberian hamsters. Hamsters were trained to run on a wheel for food pellets, in essence foraging in place for food. Their amount of foraging could then be quantified by how many pellets they obtained, as one pellet was dispensed for each 10 wheel revolutions. Siberian hamsters hoard their food rather than eat it immediately; therefore, eating did not interfere with foraging by decreasing the drive. Ghrelin administered peripherally 1 h before wheel access significantly increased this experimental foraging. Ghrelin also increased food consumption in foraging animals and animals supplied with food.

If ghrelin (or any other peripheral endocrine signaler) induces a behavioral change, it must somehow interact with the brain. In fact, ghrelin increases NPY levels in the hypothalamus. The increase in foraging caused by ghrelin can be mimicked by applying NPY to the hypothalamus and can be blocked via an NPY receptor blocker. NPY is a classic peptide neuromodulator, and, as discussed in Chapter 3, neuromodulators can cause wide-ranging changes in behavioral patterns via their adjustment of neural excitability in functional circuits. They have also been implicated in dramatic developmental shifts in behavioral profiles. The effect of the insect homolog of NPY, neuropeptide F (NPF), in *Drosophila* larvae is an example. Young larvae are attracted to food, and NPF expression is high in young larvae. As larval development proceeds, NPF is down-regulated; this coincides with food aversion, hypermobility, and cooperative burrowing. Transgenic *Drosophila* larvae with a loss of NPF show premature development of older larval behavior, whereas NPF overexpression in older larvae prolongs juvenile behaviors such as foraging while suppressing hypermobility and burrowing.

The discussion above may imply that foraging behavior is strictly driven by internal signals, but foraging is, in fact, the result of a more complicated assessment of the costs and benefits associated with the behavior. This is the foundation for optimal foraging theory, which is covered in detail later in this chapter. Although optimal foraging considers the costs and benefits of various types of food, other considerations also influence foraging behavior. One of the most immediate is the presence (or potential presence) of predators. Studies on animals as diverse as baboons and whelks show that foraging patterns are adjusted based on predation risk, and the adjustment is balanced against the severity of the animal's energy deficit. Interestingly, NPY has effects beyond eating that may play a role in this behavioral balance. It suppresses anxiety, decreases fear responses, and increases tolerance to stress. In mice, NPY-treated animals are also more tolerant of aversive stimuli, including tastes that would cause control animals to reject food. Thus, as NPY levels rise (with fasting), not only does the drive to forage increase, but the emotional responses that might tilt behavior toward caution decrease.

Caching and Hoarding

Many animal species collect food after foraging rather than consuming it immediately. Hoarding refers to collecting and storing food in home burrows. Modern humans are classic hoarding species. They forage for food in grocery stores, where they collect far more food than they can eat at one time, transport it home, and store it in dedicated areas of their home "nests" for later consumption. Many rodents, such as the hamsters discussed above, also hoard food and, unlike humans, have specialized cheek pouches that facilitate collection and transport. A more complex extension of hoarding is caching. Caching species collect food in one area, then hide it in other sites, to which they return in a kind of secondary foraging. Several species of seed-eating birds, including chickadees, tits, nuthatches, and scrub jays, cache food. Studies on caching birds are prime examples of research inspired by field animal behavior moving to a laboratory setting and then to neuroscience studies that illuminate the mechanisms behind a behavior.

Small birds live a precarious life. Their high metabolic rate dictates regular, high-calorie consumption, but food availability can vary drastically across the seasons. This environmental threat is exacerbated by low temperatures that increase metabolic demand for endotherms such as birds and mammals. In response to the selection imposed by this situation, several seed-eating bird species have evolved the ability to collect and then hide seeds in caches, to which they later return to feed. Remarkably, individual birds can cache in dozens—in some cases, hundreds if not thousands—of locations and return to them hours to weeks later. Successful caching depends critically on finding the hidden food. Field studies have shown that the individual bird hiding the food is the one who returns to eat it, and that individuals return directly to the cache rather than to a particular type of location. There is no evidence that they home in on food odors, and, in fact, they will return to a cache location even if an experimenter has removed the food from it. This suggests that birds remember the locations of their hidden food stores, even when they are numerous and scattered.

Birds can remember the location of a cache in one of two ways. One is by remembering a distinctive cue indicating the cache, similar to recognizing a street corner by remembering a particular car parked there. In nature, and on street corners, this is risky, especially when one depends on finding the location weeks later. The second is using the spatial configuration of all potential cues in the area, which is analogous to recognizing a street corner based on the arrangement of all the buildings, trees, and streets in the general vicinity. Using this strategy, the organism no longer depends on any one cue; if one particular tree is cut down, your memory tells you that you are on the correct corner, but some of its features have changed, not that you are suddenly in a new place. This is true spatial memory, as opposed to cue-based memory, and also allows one to build up an internal cognitive map to guide movement to and from the location. Animals also use cognitive maps to traverse the landscape during large-scale movements, which is the subject of the next chapter.

Understanding the mechanisms of behavior can help answer questions about the evolution of behavioral specializations. One possibility is that behavioral adaptations depend on acquiring specific, novel cognitive tools to support them. An alternative is that behavioral adaptations depend on enhancing commonly shared, general cognitive capacities, or using existing cognitive tools in a different way. If the latter is true, species with and without a behavioral specialization should differ quantitatively rather than qualitatively on laboratory tasks that depend on the cognition behind the behavior. Furthermore, if the novel behavior depends on enhancing an existing, generally held cognitive ability, one should see enhanced performance on behavioral tasks beyond the specific natural behavior of interest.

Laboratory studies of bird learning have shown that birds generally can use both cue-based learning and true spatial memory to learn the location of food rewards. Therefore, caching birds have not acquired a unique ability. But how they use spatial memory differs. Several clever experiments have pitted spatial memory against cue-based learning to discover species differences in cognitive strategies. In one example (Fig. 4.2), multiple feeder boxes are hung on a wall in an experimental

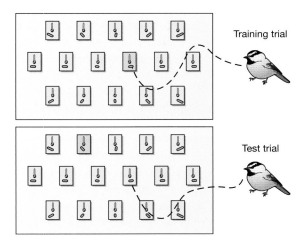

Figure 4.2. Spatial learning is demonstrated in birds. During training, food is placed in the blue box, and all birds easily learn to go to the food box. During a test trial, food is removed and the location of the blue box is switched randomly with another, previously unbaited box. Caching species preferentially go back to the previous location of the blue box rather than to the blue box itself. Noncaching species do the reverse.

aviary. All are identical brown boxes except for one that is bright blue. Birds have excellent color vision that is at least as acute as it is in humans; therefore, color differences such as this are clearly distinguishable. The blue box contains food, and the others are empty. The food is not visible until the bird removes a barrier to retrieve it. Birds are allowed to explore the feeder array and quickly learn which box contains food, demonstrating this by returning to the baited box and ignoring the others in subsequent trials. If food is removed on a test trial, birds still return to the previously baited box, showing that they remembered which was the food box (as opposed to simply using food odor for homing). But suppose on the test trial, rather than just removing the food, food was removed and the location of the brightly colored and previously baited box is switched with that of a previously unbaited box. What does the bird do? If learning is cue-based, the bird should go to the colored box regardless of its location, because it should have remembered that "bright blue = food." If spatial learning is used, the bird should go to the previous location of the bright blue box because it remembered that food was located *there* regardless of a single stimulus feature. Noncaching species do the obvious: They go directly to the blue box. Not finding food, they begin searching preferentially around the locations of previously found food. In contrast, nearly all caching species (there are exceptions) prefer to use spatial memory: They ignore the blue box and go directly to its previous location. When they find that box empty, they then search for and fly to the blue box. The search strategy for cachers is thus hierarchical, and opposite to other birds. They first prefer to use memory for location, then, if that fails, memory for discrete cues associated with the food.

Caching species also differ from noncaching birds in another way. They are better at spatial memory tasks in general. In a variety of laboratory tasks, they learn locations more quickly and remember them better. The most consistent difference between cachers and noncachers is that spatial memory duration is superior in the caching species. This trait is crucial for retrieving hidden food and suggests that strong selection based on behavioral ecology has resulted in changes in cognitive strategy and enhancement of particular memory processes.

Independent of work on natural animal behavior, memory mechanisms have long been the subject of biomedically oriented neuroscience. The compelling case of a patient known as H.M. focused attention on the hippocampus as a brain area crucial for memory formation. H.M. had severe memory impairments after bilateral hippocampal removal to alleviate severe epilepsy. His simple stimulus–response, or cue-based, learning was spared, but the ability to form and use more complicated cognitive memories related to rules and relationships, contexts, social episodes, and, important here, spatial locations, was severely compromised.

Spatial learning has been adopted as a standard model for neuroscience research on complex learning in rats and mice, and hippocampal plasticity remains a major topic of neuroscience research spanning molecular to behavioral studies. Is the evolution of caching and the enhanced spatial memory that supports it reflected in

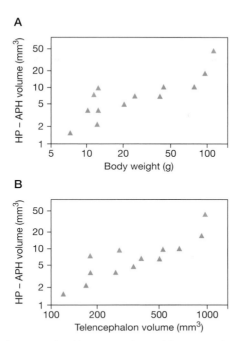

Figure 4.3. Relationship between the hippocampal–parahippocampal area (HP-APH) and body weight (A) and telencephalon volume (B) in passerine subfamilies that do (blue triangles) or do not (red triangles) cache food. The allometric relationships show that caching birds have, on average, larger-than-expected hippocampal complexes than noncaching birds for their body size and a larger-than-expected portion of their telencephalon represented by the hippocampal complex.

the hippocampus? The answer is yes: Caching birds, compared to noncaching congeners, have a larger hippocampus (corrected for overall brain or body size; Fig. 4.3). Thus, the increase in spatial memory abilities necessary for caching is reflected in an increase in the size, and presumably neural processing capacity, of the brain area responsible for spatial memory.

Good spatial working memory is important in other foraging domains. Consider active foragers, which must travel away from a home site and find their way back, versus sit-and-wait predators, which remain at home and catch food in an opportunistic fashion. Lainy Day examined the spatial learning abilities and relative hippocampal size in two congeneric lizards with these opposing feeding strategies, *Acanthodactylus boskianus*, an active forager that seeks out clumped sedentary prey, and *Acanthodactylus scutellatus*, a sit-and-wait predator. As with cachers versus noncachers, the active foragers had superior spatial learning in a laboratory task (remembering the location of hidden hot rocks in a cold sand arena) and a larger hippocampus.

Caching birds and actively foraging lizards tell us something important about the evolution of behavioral diversity, and whether changes in cognition that accompany new behaviors are domain general or domain specific. Specialized behavior depends on the enhancement of generalized cognitive abilities rather than on the addition of new cognitive functions. The enhancement is reflected in behavioral domains beyond the specific task we are focused on. And finally, the brain is remodeled in predictable ways, most noticeably in expansion of particular brain areas to accommodate the enhanced cognitive abilities underlying the behavioral adaptation.

PREY CAPTURE: THE CONSUMMATORY PHASE

Once found, a food item is taken and eaten. Ethologists considered this aspect of feeding stereotyped and stimulus triggered, although, like the foraging that precedes it, gated by physiological drives. In many species, prey capture does indeed have these qualities, although not to the extreme of earlier views. Prey capture in toads, barn owls, and pit vipers are three examples of consummatory behavior, in each case triggered by a different sensory cue. They are also examples where animal behavior studies stimulated research into the mechanisms of natural behavior, which, in turn, revealed fundamental insights into how the vertebrate brain processes information.

Visual Prey Capture in Toads

Common toads (genus *Bufo*) are foragers that capture small insects or worms. Toads rapidly orient toward potential prey, then lunge and snap with stereotyped and very rapid tongue and jaw movements. The behavioral sequence is triggered by visual stimuli with particular size, shape, and movement characteristics. Behavioral analysis shows that the stimuli can be easily and simply defined so that simple rectangles moving in the direction of the long axis are perfectly good triggers of the prey capture response (Fig. 4.4). A nonsense stimulus (from a natural perspective) such as a rectangle moving perpendicularly to its long axis, movement of which no real worm would be capable, triggers nothing or an avoidance response. Other visual stimuli also trigger stereotyped predator avoidance behavior. When faced with one of their natural predators—a snake or large bird or mammal—toads orient away from the threat to present the flank (rather than the head), body inflation, and a stiff-legged posture, all of which turn the toad in an escape direction and increase its apparent size. As with prey capture, any large moving object will elicit the response. In short, toads use simple visual features that abstract key elements of prey and predators and use those to rapidly trigger the appropriate reflex reaction. Although details of the movement sequences are somewhat variable depending on factors such as distance to the stimulus, toad prey capture and predator avoidance are as close to traditionally defined sign-stimulus-evoked fixed action patterns as natural behavior gets.

In 1959, Jerry Lettvin and colleagues asked a simple question that became the title of an article that probably started the field of neuroethology: What does the frog's eye tell the frog's brain? This question had never been asked by neuroscientists in

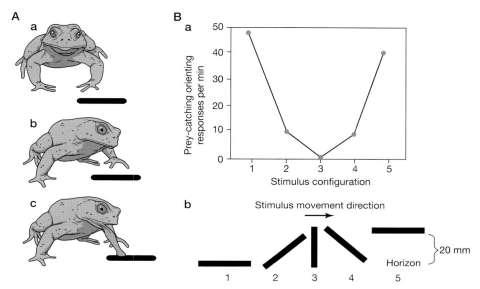

Figure 4.4. (A) Toads orient (a), lunge (b), and extend their tongue and snap a small, moving stimuli (c), whether a worm or a dark rectangle on a white background. (B) Responses (a) depend on the direction of movement relative to the long axis of the stimulus. (b) The orientation of the stimuli indicated on the x-axis of the graph in Ba.

quite the same way. What Lettvin was asking was, how does the eye communicate to the brain the environmental features that guide natural behavior? In subsequent years, work by his lab, David Ingle, Jörg-Peter Ewert, Ursula Grüsser-Cornehls, and others answered that question and unraveled the brain systems responsible for visually guided behavior in anurans. Electrophysiological recordings of visually evoked responses in retinal ganglion cells, the cells in the eye that transmit information from the retina to the brain, show a variety of cell types. At least four only respond to moving objects (a key sign for both behaviors), as long as the stimuli have sufficient contrast and are within an appropriate size range. The toad's retinal output cells are more specialized for movement detection than are those of many other vertebrates (e.g., mammals) and are clustered in their size preferences around the sizes of naturally important objects. Hence the eye has properties that are skewed toward detecting objects that are ecologically meaningful to the organism. The retinal cells alone, however, are not sufficient to predict the behavioral responses, particularly the decision (based on stimulus features) about whether to approach or avoid. For this task the brain must therefore build up more specialized feature detectors and decision networks.

The circuit underlying frog prey capture is shown in Figure 4.5. The retinal ganglion cells send their outputs to several brain regions, most important for this story

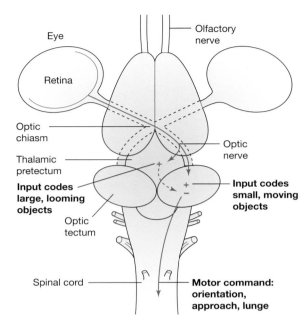

Figure 4.5. Schematic drawing of a frog brain seen from the top, with the forebrain toward the *top*. Retinal inputs terminate in parallel in the pretectum and optic tectum. Retinal inputs to the pretectum code large, looming objects and stimulate cells there; tectal inputs to the optic tectum code small, moving objects and stimulate cells there. Tectal cells send outputs to spinal cord motor areas and stimulate orientation, approach, and lunging movements. Pretectal cells send outputs to the optic tectum and inhibit cells there. Thus, when a large object is seen, the tectum's orientation and approach movements are inhibited.

to the optic tectum, a large midbrain structure (called the superior colliculus in mammals), and to the pretectum (a collection of nuclei in the caudal diencephalon just in front of the colliculus). In both cases, the connections from the eye terminate in a "retinotopic map," an organization that preserves the spatial arrangement of the retina and therefore represents a map, or topographically conserved representation, of the visual world. Several different types of tectal neurons exist, but most relevant to natural behavior is the class of neurons termed T5 cells. T5 cells are more selective in their responses than are retinal cells. They are strongly sensitive to object movement and to configuration, and they have clear preferences for objects shaped like narrow rectangles. The T5(2) subclass cells respond to objects moving in parallel to the long axis of the rectangle, but not when the same object moves perpendicularly to the long axis. They, in essence, are "worm detectors" that closely mirror the behavioral preferences in their electrophysiological response. T5(2) cells connect directly to brainstem motor areas that mediate reflexive behavioral responses.

Pretectal cells are also more selective than retinal cells, but in the opposite behavioral direction to tectal cells. Pretectal TP3 cells respond vigorously to large, high-contrast objects. Moreover, pretectal cells connect to tectal cells, where they inhibit tectal cell responses. The connections maintain the retinal mapping, so that pretectal cells monitoring a particular area of the visual world connect with tectal cells monitoring the same area (both via parallel inputs from the retina). Tectal and pretectal cells are using those parallel inputs differently, however; the tectal cells use the information to construct "prey detectors," whereas the pretectal cells use it to detect "threat detectors." The pretectal–tectal connections represent the most basic decision circuit encountered in the brain. When a small, moving, worm-like object is detected, the pretectum is silent and the tectum triggers prey capture. However, when a large, looming object is detected by the pretectum, it quickly inhibits the tectal cells that would trigger an orientation–approach–lunge response. As predicted, David Ingle found that lesions of the pretectum cause frogs to orientate and lunge at objects that normally would be viewed as a potential threat.

Auditory Hunting by Owls

Like toads, barn owls (*Tyto alba*) are nocturnal hunters. But they rely on sound rather than vision (although they will use their excellent eyesight if possible), and fly down on small rodents, grabbing them with their talons, rather than hunting from the ground and lunging to grasp prey with jaws and tongue. At first, hunting by barn owls may seem very different from prey capture by toads, and, in fact, it is more complicated. However, there are elements that illuminate common mechanistic principles of feature detection related to consummatory behavior.

Small rodents such as field mice are a major part of the barn owl's diet. Detecting mice from a distance and localizing these moving targets, which are often running in or under leaf litter, accurately enough to swoop down for successful capture is a daunting task. Remarkably, barn owls can achieve this in complete darkness using sound alone. Laboratory experiments show that sound localization in barn owls is very accurate in both the horizontal and vertical directions (very unusual in vertebrates, but something important when trying to triangulate a prey's sound from above). In laboratory experiments, their error range from a distance of 5 m is only 8 cm, which is about the size of a mouse. Although the sensory processing responsible for auditory localization is more complicated than for the visual specializations in toads, both systems depend on setting up a peripheral sensory system attuned to behaviorally relevant cues, then processing in the central nervous system (CNS) to construct feature detectors.

One specialization of the barn owl ear is that frequencies from about 5000 to 9000 Hz are overrepresented (barn owls can hear frequencies from 100 to 12,000 Hz, or about the lower half of the human hearing range). Sound localization accuracy varies by sound frequency depending on a complicated interaction of several stimulus and receiver characteristics, but modeling suggests that these relatively high frequencies provide the best sound localization potential in birds the size of owls. In vertebrates, in general, sound localization depends on comparing between the two ears time and intensity differences, which vary together depending on the relative location of the sound in the horizontal plane around the animal's head, or azimuth.

Owls too use this basic strategy, but a remarkable physical specialization of the face enhances sound localization and makes possible the accurate two-dimensional calculations underlying successful prey strikes (Fig. 4.6). The face and feather ruff around each ear are shaped like twin parabolic dishes, similar to the plastic background dishes used to collect sound and enhance directional sensitivity in microphones. Furthermore, the two ears are asymmetrical. The left ear opening is above the midpoint of the eyes and points downward, whereas the right ear does the opposite. This induces an asymmetry in hearing not found in other vertebrates. The left ear is more sensitive to sound coming from below, and the right ear more sensitive to sounds coming from above, the midpoint of the owl's face. Thus, an owl can compare intensity differences from the two ears to calculate the elevation of a sound source. At the same time, it can compare the time of arrival at the two ears to calculate the position of a sound source in the horizontal plane. Together, the sound can be placed in elevation and azimuth to guide the owl's strike.

Vertebrate ears do not share information directly, so any comparison of the inputs to the two ears must be done in the brain. Anatomically, and in many physiological features, the owl brainstem areas of the auditory system are typical of birds and of other vertebrates. At lower processing levels, inputs from the two ears converge onto neurons so that interaural comparisons can be made and represented in the

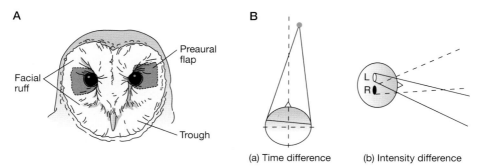

Figure 4.6. (A) The facial anatomy of barn owls is specialized to enhance vertical and horizontal sound localization. The ears are asymmetrical and the facial structure and facial ruff feathers form parabolic dishes around the ear canals. (B) The facial specializations allow barn owls to determine (a) the horizontal position of a sound by comparing interaural time differences based on the separation of the ears and (b) the vertical position of a sound based on interaural intensity differences created by the vertically asymmetrical ears. The sound can then be pinpointed when the midbrain places the stimulus on a sound location map with two axes, azimuth (horizontal plane) and elevation (vertical plane), relative to the owl's face.

activity of those brain neurons. As is typical, time comparisons (here azimuth) and intensity comparisons (here elevation) are kept separate. These ultimately converge in the midbrain, in a nucleus that is part of the inferior colliculus. There individual neurons can calculate, based on both time and intensity, the true location of the sound. They in essence become feature detectors for particular sound locations.

But how does the owl's brain "know" what interaural difference corresponds to a location in the real world? In a series of several elegant studies, Eric Knudsen discovered that the auditory system depends on the visual system, which intrinsically codes real location, to do this. The auditory space information is sent to the optic tectum, or superior colliculus, which has a visual map of the world, and forms an acoustic–spatial map in register with it (Fig. 4.7). The experience of seeing and hearing a sound-producing stimulus during critical phases of juvenile development is crucial to the owl's ability to hone the mapping and put the visual and auditory spatial representations in register. Note that the optic tectum is the same structure used by toads to link the visual representation of the world with a motor response to a stimulus's location. The owl's optic tectum (and the tectum of every other vertebrate) does the same thing. The enhanced acoustic–spatial map, once properly organized through behavioral experience, can piggyback onto this system so that in the end either the sight *or* the sound of a mouse can direct the rapid head turns and movements toward it with great accuracy.

Infrared Prey Sensing by Snakes

When a new sensory adaptation for hunting arises, the old rules are still followed. Infrared (IR) detection has evolved independently in two groups of nocturnally hunting snakes, crotalines (pit vipers) and boids (constrictors). IR sensing is detecting and locating heat at a distance, and some snakes can use it for finding heat sources such as prey (small mammals), as well as warm-blooded predators or thermal refuges. IR sensitivity in snakes is quite high—they can detect the equivalent of a $0.001\,°C$ difference between a target and the ambient temperature. Although we normally do not think of detecting heat locations accurately, pit vipers will orient toward a heat source with an accuracy of between $2.5°$ and $5°$ of visual angle, more than sufficient to guide a strike at a rat from a meter or two away. Such strikes are, like frog prey-catching and owl orientation, very fast and relatively stereotyped. Snakes use a variety of other sensory stimuli, including vision, smell, and vibrations, to find prey, but pit vipers can orient and strike prey based on the heat signature alone. It is as though the snake sees and is guided by an infrared image of the object. But pit vipers and boa constrictors do not use the eye to see into the infrared range. Instead, they have evolved specialized pit organs on the face that serve as thermal detectors. These organs are the reason that pit vipers like rattlesnakes have this name.

As any evolutionary biologist might suspect, pit organs are derived from a sensory apparatus found in other snakes and in vertebrates in general—the trigeminal temperature nerve endings found in the skin of the face and jaw of all vertebrates. A few

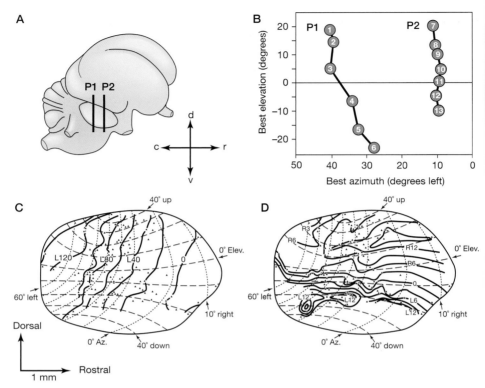

Figure 4.7. Neural mapping of sound location in the optic tectum of the barn owl. (*A*) Drawing of barn owl brain (side view) showing location of two electrode penetrations. Arrows indicate anatomical direction (<u>d</u>orsal, <u>v</u>entral, <u>c</u>audal, <u>r</u>ostral). (*B*) Elevation and azimuth tuning (relative to center of the owl's face) of cells encountered in the two electrode penetrations illustrated in *A*. Cells in electrode track P1 change their elevation sensitivity, but maintain relatively constant azimuth sensitivity as the electrode moves dorsal to ventral. Cells in track P2, located more rostrally, jump to a different best azimuth, and again change their elevation sensitivity from dorsal to ventral in the tectum. (*C*,*D*) Maps of best interaural time difference (ITD; perceptually corresponding to azimuth, in 20-μsec intervals; e.g., "L120" means the cells along that line respond best when sound hits the left ear 120 μsec before the right ear) and best interaural intensity difference (IID; perceptually corresponding to elevation, in 3-dB intervals; e.g., "R6" means cells along that line respond best when sound is 6 dB louder in the right ear) across the tectal surface. Dorsal and rostral directions are indicated by the arrows. The ITD and IID lines are oriented perpendicular to each other to form an acoustic space map such that sounds of a particular azimuth (ITD value) and elevation (IID value) indicating a specific location in front of the owl are coded by cells on a particular spot in the optic tectum. The auditory space map is in register with the tectum's visual map.

hundred trigeminal nerve endings similar in structure to skin temperature receptors line the back of each shallow pit. These neurons connect with an expanded area of the descending trigeminal nucleus in the brainstem, the sensory nucleus that in all vertebrates processes skin temperature and pain information. Thus, snakes have evolved specialized IR "eyes" from the generalized temperature-detecting skin receptors of vertebrates (Fig. 4.8A,B). But it is a very poor eye based on anatomical and functional constraints. Not being a true eye, but a skin indentation to concentrate existing temperature receptors, it has no lens to focus the IR image and thus sharpen its spatial resolution. Instead, the pit functions as a pinhole camera, using the opening

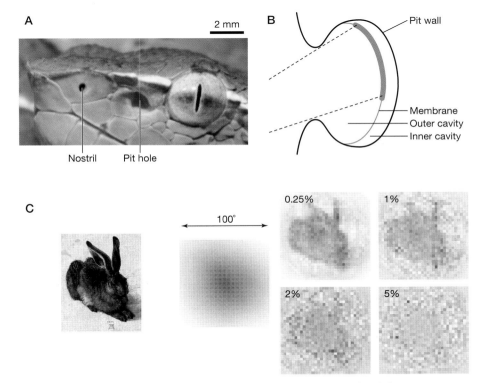

Figure 4.8. (A) The infrared-sensing pit organ is located between the nostril and the eye in pit vipers. (B) The organ is a simple indentation with temperature-sensitive nerve endings lining an inner membrane. There is no lens or complicated neural network in the pit organ itself. (C) A model of central processing, based on simple, standard operations such as lateral inhibition, can radically enhance the resolution of the thermal image coded by the pit organ. As an example, the modeled thermal signal of a rabbit (*left*) shows the image "seen" by the pit (*middle*) to be extremely ambiguous. The neural network model applied to the image can enhance the image in the brain, with resolution depending on signal noise from object movement, environmental interference, or receptor noise. The *right* four panels show the image when cumulative noise is 0.25%, 1%, 2%, and 5% of the maximal intensity of the stimulus on the pit membrane.

of the pit to provide a focused projection onto the pit's sensory lining. To increase sensitivity, however, the pit must be open as wide as possible to allow as much IR radiation as possible to enter. The wider the opening, the greater is the sensitivity (necessary for distance detection), but the poorer the focus, which the organ needs to detect the edges of the object precisely.

As for both the toads and the owls, the brain needs to use limited sensory information coming from the peripheral sense organs to construct a more accurate representation of the image. It is known from electrophysiological studies of the IR system that the poor spatial resolution seen in the pit receptors progressively improves at each successive stage of IR processing in the brain. The neural circuitry underlying IR processing is not as well understood as are the circuits for frog visual prey capture or owl sound localization. However, computational models based on physical characteristics of the pits, heat distributions, and common neural operations such as lateral inhibition (common in many sensory systems for sharpening neurons' detection of edges and boundaries) clearly show that the poor image from the pit can be drastically sharpened without invoking anything other than the neural circuits expected to be present in any low-stage central sensory processing centers (Fig. 4.8C).

By the time IR information reaches the midbrain, spatial resolution of the IR coding neurons is much better than in the pit receptors and is organized into a thermal map consisting of neurons that can resolve the location of an IR source with approximately the same accuracy as occurs in a snake's behavior. This map is in the optic tectum, in register with the visual map seen there in toads, owls, and all other vertebrates. Just as the constructed sound-localization map of owls can employ the existing visuomotor connections to guide its sonic prey orientation, the IR map in pit vipers can use the same visuomotor connections to guide thermal prey strikes.

Common Patterns in the Sensory Basis of Prey Capture

All three cases discussed above show how the relatively stereotyped consummatory phase of prey capture depends on sensory triggers that represent important features of the prey and its location. The evolution of these specialized behaviors depends at least in part on the evolution of specialized sensory characteristics that, at their root, are expansions or enhancements of existing sensory systems or sensory motor interfaces such as the optic tectum found in all vertebrates. Sense organ coding is never enough to truly explain the behavior, however. Instead, circuitry of the CNS is necessary to compile and synthesize the input into progressively more narrowly tuned neurons that serve as detectors for key features of the prey item or its location. This view of natural sensory processing, based on understanding the behavior of diverse animals adapted for specialized tasks, now suffuses sensory neuroscience in general, where key questions guiding research in that field are based on stimulus features that are ecologically relevant for detecting real-world objects, and understanding how these features are extracted and represented in the brain.

FOOD PREFERENCES

Every biologist knows that species are associated with different kinds and ranges of food preferences, from very specialized to very general, and everyone who has owned a pet knows that individual animals vary in their preferences within the species type. In a few cases, the genetic correlates of food preferences or tolerances are known. These are usually related to genes coding olfactory or chemical receptors that allow one group to detect or fail to detect chemical signatures. Two sister species of *Drosophila*, *Drosophila simulans* and *Drosophila sechellia*, are, respectively, generalist and extreme-specialist foragers. They differ significantly in their genetic makeup related to chemoreceptor proteins expressed in the antennae (see Chapter 3). Genetic differences would likely set some basic feeding parameters. In addition to this, behavioral experience with food can have dramatic effects on future food preferences.

Animals clearly learn to avoid the smell, taste, or sight of food that was aversive or caused an illness. Such taste aversion learning has long been interesting to comparative psychologists because its features seem very different from traditional conditioning. If the aversive reaction is strong enough, one encounter is enough to cause very powerful, long-lasting learning. Such powerful one-trial learning is rare. The aversive cue, such as becoming sick, can occur long after the food was ingested, but still support the learning as a negative reinforcer. Generally, a reinforcer must be closely associated in time with a cue for the cue to become associated with the event. Last, taste aversion learning has a strong affective and visceral component. A human who acquires a taste aversion not only knows to avoid the food, but can feel physically ill and revolted by its smell or sight. Rats must feel the same way. When rats first taste a noxious substance or are sickened by something they eat, they demonstrate face wiping and other aversion behaviors. When they later simply smell an odor associated with experimentally induced taste aversion, rats show the same face wiping, gaping, and other aversion behaviors.

Food aversion is powerful, but not necessarily permanent. What is learned can be unlearned given sufficient experience, and this could be adaptive if the learned aversion was based on a unique event rather than something intrinsic to a class of foods. For example, an animal may become ill by eating a single contaminated grape, but should not then avoid anything that smells like a grape. But how would an animal know to try a food item again? Surprisingly, observational learning has been shown to overcome learned taste aversions in rats. Rats that observe other rats eating food they previously found unpalatable are more likely to sample the food again. In fact, social facilitation of food preferences or for trying unfamiliar foods has been demonstrated in species as diverse as rats, bats, spotted hyenas, and red winged blackbirds, and cultural transmission of food handling preferences has been seen in primates and bats. We noted in Chapter 3 that mate choice is also influenced by observing conspecifics' choices, a phenomenon called "mate copying." Food preference modulation by observing the choices of conspecifics is a fundamentally similar behavioral mechanism controlling what one might first assume is a fixed trait.

Preferences and Defenses

Lessening food aversions only works if the subsequent encounter is less aversive than the first. In many cases, potential prey have developed defense mechanisms that ensure this will not happen. Noxious tastes, or in the extreme, life-threatening toxins, have evolved many times by prey species to thwart predation. Organisms that defend themselves in this way are often adorned by bright colors or distinctive smells or tastes that advertise their offensive nature. Manipulating taste preferences underlies the evolution of such aposematic signaling.

Newts in the genus *Taricha* are only one group that has evolved a particularly deadly neurotoxin as a predator defense. Tetrodotoxin blocks sodium channels in neurons and skeletal muscles. Movement of sodium ions through neuronal membranes is the basis of neural signaling, and when sodium channels are blocked, nerves and muscles cease to function. Extremely small doses cause numbness, tingling, and motor impairment; slightly larger doses cause death. If an animal tries to eat one of these newts, the unpleasant feelings in the mouth generally cause it to spit out the newt. If it happens to eat it anyway, serious debilitation will occur because a *Taricha* newt contains enough tetrodotoxin to kill an adult human. Like many poisonous organisms, *Taricha* have aposematic coloring. They have bright-colored bellies, which they display in stereotyped postures when threatened (Fig. 4.9). Several other species of newts, as well as pufferfish, and a few other marine organisms, have independently evolved the ability to concentrate lethal levels of tetrodotoxin. Like *Taricha*, they face a problem adopting this strategy: How do they avoid killing themselves? Sensitivity to tetrodotoxin is determined by the presence of an aromatic amino acid in the protein making up the outer potion of domain 1 of the sodium channel, where the toxin binds and blocks the channel. All animals that concentrate tetrodotoxin as a chemical defense have convergently evolved a non-aromatic amino acid substitution in the tetrodotoxin-sensitive domain, resulting in a sodium channel that has limited binding to the toxin. This keeps them from poisoning themselves.

The relationship between predator and prey has been likened to an arms race, and the example of *Taricha* demonstrates this. Although nearly everything else avoids these newts, *Taricha* are a major food source of aquatic garter snakes (*Thamnophis sirtalis*). *Thamnophis* has, evolutionarily speaking, responded to the newts in kind. To combat the threat of the toxin, garter snake sodium channels show the same domain 1 gene variants that have evolved to keep the newts themselves alive. The coevolution of predator and prey is clearly seen in studies by Edmund D. Brodie III on geographic variation in tetrodotoxin resistance among garter snake populations. Resistance varies and is absent in garter snake populations that do not overlap with toxic newts (Fig. 4.9). Mitochondrial DNA analysis shows that the resistant populations

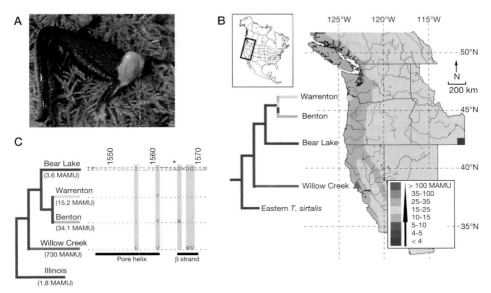

Figure 4.9. (A) *Taricha* spp. Newts, showing aposmatic coloring when threatened, concentrate the neurotoxin tetrodotoxin as a predator defense. (B) Populations of garter snakes (named by location, with their genetic relationships shown by the cladogram) have evolved resistance to the toxin where they overlap with toxic newts (newt distribution shown by shading). Resistance is color-coded by mass-adjusted mouse units (MAMU), a measure of the dose of toxin needed to cause a standard level of motor impairment. (C) The amino acid substitutions in domain 1 of the sodium channel in garter snake populations. The Bear Lake population is not toxin-resistant. Resistant populations have independently evolved substitutions that decrease tetrodotoxin binding to the sodium channel.

independently and convergently evolved similar substitutions in the same coding region for the domain 1 portion of the sodium channel gene.

The influence of tetrodotoxin defense on the evolution of resistance can be seen in other animals. Pufferfish have it both ways. In addition to having tetrodotoxin as an effective predator defense, their tetrodotoxin resistance allows them to feed on toxic dinoflagellates avoided by nearly all other species. Where dinoflagellates producing these tetrodotoxin-like toxins are in high concentration, softshell clams have evolved a similar toxin-resistant sodium channel. These simple gene variants represent an unusual genetic mechanism underlying the evolution of a food preference that allows exploitation of a food source with little competition from other species.

EXPLOITATIVE FORAGING: PARASITE-INDUCED BEHAVIOR

Strange as it may seem, in some cases being eaten is an adaptive strategy. Many parasites have complex life histories that require residences in sequential host species during the parasite's life cycle. The problem that such parasites face is how to leave the current host and enter the future host. One solution is to get your current host, and yourself along with it, eaten by the future host. Surprisingly, parasites can manipulate host behavior to increase the possibility of that happening.

One could argue that simply compromising a host's health, making it less able to escape from predators, is a parasite-induced change in host behavior that benefits the parasite by increasing its chances of moving to its next host. Far more interesting, however, are examples of specific behavioral changes unrelated to the compromised condition that coincides with parasitic infection. The trematode *Dicrocoelium dendriticum* is one example (Fig. 4.10). Trematodes infect the guts of sheep and cows as liver flukes. Mature trematodes secrete eggs in the herbivore's feces, where land snails pick up their infection as the first intermediate host. Juvenile trematodes are discharged in the snail's slime and are then picked up by ants, their second intermediate host, which use the slime as a moisture source. Most juvenile parasites infest the ants' hemocoel, the spaces between the organs of most arthropods and mollusks through which the hemolymph circulates, but at least one will enter and reside in the ant's nervous system, specifically the subesophageal ganglion. Ants infected with

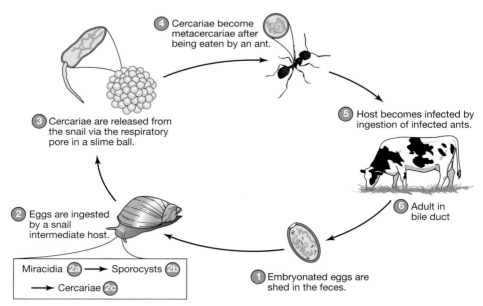

Figure 4.10. Life cycle of a trematode parasite. Ants infected by the trematode exhibit changes in behavior that stimulate them to move to the top of grasses and hence be exposed to being eaten by grazing herbivores. Herbivores are infected by the trematode, whose eggs are shed in the herbivore's feces. The eggs are ingested by snails, where the eggs develop into juveniles, which are secreted in the snail's slime deposits. Ants are infected when they use the slime as a moisture source and eat the juveniles. Juveniles mature in the infected ants, completing the parasite's life cycle.

D. dendriticum demonstrate a very odd behavior, but otherwise show no sign of compromised health. At night, triggered by cooling temperatures, they move away from their colony, climb to the tops of grasses, clamp on with their mandibles, and sit through the night and into the morning. The ants and their parasite companions then become prey to herbivores grazing in the morning on the grass, completing the infection cycle. If they do not become fodder for the grazers by the time temperatures rise enough to dry out and kill them, the ants release their grip, descend the grass, and rejoin the colony. As yet, the mechanism for this astounding host behavioral manipulation is unknown.

An equally compelling example is provided by *Polymorphus paradoxus*. Like many other members of Acanthocephala, these small parasitic worms infest the gut of birds, specifically ducks. Eggs are secreted in the feces and then eaten by small aquatic crustaceans, here *Gammarus lacustris*, which serves as an obligate intermediate host. The eggs hatch and develop into an intermediate infectious stage, the cystacanth, which lodges in the crustacean's hemocoel. Ducks eat the crustaceans and become the terminal host, restarting the cycle. Infected *G. lacustris* radically alter their escape behavior to make it more likely that they will be eaten. They switch from being photophobic to photophilic, causing them to swim to the surface rather than to the bottom of a pond when startled. Infected animals also swim in a straight line along the surface until they encounter a solid surface, at which time they cling to it. All of this makes it more likely a duck will encounter and eat them. Infected crustaceans are, in fact, far more likely than uninfected to be caught and eaten by ducks and other waterfowl. *P. paradoxus* and its congeners, which induce similar specific behavioral changes, do not contact the crustacean's nervous system. However, by an as-yet-unknown mechanism, they induce a significant increase in the crustacean's brain serotonin levels and morphological changes in serotonin neurons. As noted in Chapter 3, serotonin is an important neuromodulator in invertebrates (as well as vertebrates), and manipulations of serotonin levels can drastically change activity in neural circuits, including the central pattern generators underlying locomotion and escape responses. Experimentally elevating serotonin in uninfected *Gammarus* induces behavioral changes identical to parasite infection.

DECISIONS ABOUT FOOD: OPTIMAL FORAGING

Animals not only have to consider what food is good for them but whether it is worth pursuing. What should an animal do when faced with a choice of searching for a high-quality, but rare, food versus taking close and more abundant low-quality food? What factors might enter into its decision about how to proceed? And, theoretical demonstrations aside, can and do animals actually vary their foraging behaviors in response to complicated trade-offs of quality, effort, and risk? This question of how foraging decisions might be modulated by the relative costs and benefits of acquiring its food gave rise to the field of optimal foraging, introduced by Robert H. MacArthur, Eric R. Pianka, and J. Merritt Emlen, as one of the first explicit integrations of ecology and behavior. It is also one of the fields of animal behavior in which detailed mathematical models can be seamlessly linked to experimental tests of predictions they make.

Optimal foraging theory assumes that an animal is an optimal forager, and the theory makes precise mathematical predictions as to what decisions the animal makes about prey choice. The predictions can be experimentally verified, and when the predictions are not met, they can lead to subsequent studies that reveal cognitive constraints as to why an animal is not foraging optimally.

An Optimal Foraging Model

The most fundamental foraging decision an animal makes is whether to eat or not to eat something. Intuition suggests that this decision should be based on the value of the food item. The basic optimal foraging model asks if an animal should feed on all prey it encounters or if it should reject the less profitable prey. In other words, when should an animal be a generalist or a specialist?

The optimal foraging model makes two crucial assumptions: Prey vary in their profitability, and predators behave in a manner that maximizes their fitness. Profitability is determined both by the energy that is derived from the prey (E) and by the costs of obtaining it (T). The fitness of foragers is usually not measured directly, but instead a proxy for fitness is used. The proxy is the benefit-to-cost ratio associated with foraging, E/T, and the optimal forager is the one that maximizes this ratio. The model also assumes that the forager can ascertain the profitability of the prey items and knows the encounter rate, and that the forager cannot search while the prey item is being handled.

In the basic model, there are two prey types available to a forager, and the payoffs of being a specialist, preying on only one prey type, are contrasted with the payoffs of being a generalist, preying on both prey types. Prey vary in the energy they contain (E), how often they are encountered (λ), and the time it takes to subdue the prey or its handling time (H). Predators search for prey for a certain amount of time (S) and encounter the prey randomly during searching at a rate (λ) that varies with the prey's availability. This simple set of variables allows one to calculate the relative costs and benefits of the two foraging strategies, specialist versus generalist.

The costs of foraging can be calculated as

$$T = S + S(\lambda_1 H_1 + \lambda_2 H_2)$$

for a generalist preying on two prey types, and more simply as

$$T = S + S(\lambda_1 H_1)$$

for a forager specializing on only one prey type. This is simply the amount of time the forager with each strategy devotes to searching for prey. Similarly, the benefits of foraging for a generalist is

$$E = S(\lambda_1 E_1 + \lambda_2 E_2)$$

and for a specialist is

$$E = S(\lambda_1 E_1).$$

This is an estimate of the energy intake obtained while foraging under each strategy.

The fitness of the forager, E/T, can be calculated for the generalist (gen) and the specialist (spec):

$$E/T_{\text{gen}} = S(\lambda_1 E_1 + \lambda_2 E_2)/S + S(\lambda_1 H_1 + \lambda_2 H_2),$$

while

$$E/T_{\text{spec}} = S(\lambda_1 E_1)/S + S(\lambda_1 H_1).$$

The model predicts that a forager should be a specialist instead of a generalist when its payoffs are greater, when $E/T_{\text{spec}} > E/T_{\text{gen}}$, or more specifically, when

$$S(\lambda_1 E_1)/S + S(\lambda_1 H_1) > S(\lambda_1 E_1 + \lambda_2 E_2)/S + S(\lambda_1 H_1 + \lambda_2 H_2).$$

This latter inequality reduces to

$$(E_1 H_2/E_2) - H_1 > 1/\lambda_1.$$

This last inequality makes some very simple and some unintuitive predictions about how animals should forage. A predator will be more likely to forage as a specialist when the encounter rate with the more profitable prey (λ_1), prey 1, is higher and thus $1/\lambda_1$ is lower. Inversely, a forager is more likely to be a generalist when the encounter rate with the more profitable prey is lower. These are predictions that are fairly intuitive: When the profitable prey becomes less common, then the forager should not specialize on it. But an unintuitive prediction is that the relative abundance of the two prey types should not figure into the animal's foraging decision, only the absolute abundance of the more profitable prey. Even if the less profitable prey is much more abundant than the more profitable one, the forager should be a specialist as long as the encounter rate of the more profitable prey is above some threshold.

Testing the Model

Mathematical models are important in behavior because they force the researcher to acknowledge all the variables that come into play, recognize the interactions among them, and ultimately make testable predictions. The internal validity of a model can be tested by making sure the mathematics is correct. The external validity, whether the model is a useful biological tool, must be tested in the laboratory or the field. How can one test the predictions made by this optimal foraging model?

John Krebs and his colleagues studied foraging decisions in a common European bird, the Great Tit, using a cleverly simple device that allowed them to vary the relevant parameters necessary for testing optimal foraging theory. A great tit was perched above a conveyor belt on which it could view mealworms passing below. The mealworms were of two sizes and thus varied in the energy (*E*) the bird obtained from them, but the handling time for each prey type (*H*) did not vary. Once a bird took a mealworm, it flew back to its perch to eat it; thus, the birds did not search while handling their prey. The encounter rate (λ) and the absolute abundance of each prey were varied. In this way the researchers could test the prediction that the foraging decision of being a specialist or a generalist was determined by the overall encounter rate with the more profitable prey and not the relative abundance of the two prey items.

In the first treatment, the densities of two prey were low and the prey were present in equal proportions. The optimal foraging model predicts that the bird should behave as a generalist under these conditions, and that is what it does, taking each of the prey items in equal proportions (Fig. 4.11). In the next treatment, the density was increased and the encounter rate of the more profitable prey was increased relative to that of the less profitable prey. As predicted, the bird now shifts to a specialist strategy choosing the larger prey item almost exclusively (Fig. 4.11). But can one be sure it is

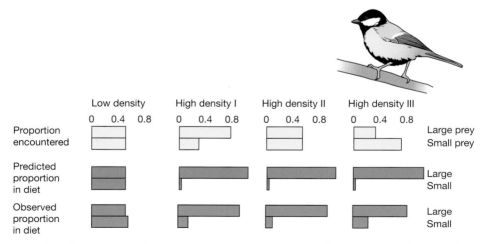

Figure 4.11. Results of experiments conducted on great tit foraging behavior. Researchers varied the density of two prey items as well as their relative proportions. For each treatment they predicted whether the forager should be a generalist (i.e., take each prey item in proportion to their encounter) or if they should specialize on the most profitable prey item. The results support the model showing that the absolute abundance of the less profitable prey does not influence the foraging decision.

the encounter rate with the profitable prey and not the relative density of the two prey that is the causative variable? If the experiment is repeated at the same density, but now the encounter rate of each prey item is equal, the bird still behaves as predicted and acts like a specialist (Fig. 4.11). The final treatment generates the most unintuitive results, which support the most unintuitive prediction of the model: Only the encounter rate with the more profitable prey matters. If the high density of prey is maintained but the encounter rate of the less profitable prey is now higher than the encounter rate for the more profitable prey, the Great Tits still strongly bias their choice of prey to the more profitable one (Fig. 4.11).

Mathematical models of behavior are more common in behavioral ecology than in other areas of animal behavior. At their best, these models do not merely formalize verbal arguments, but they also make predictions that might not result without the mathematics. But internally elegant models might do little to advance our understanding of nature unless their external validity can be tested. Optimal foraging might represent the most powerful union of modeling and experimentation in animal behavior.

Assumptions about Learning

One of the assumptions of the optimal foraging model is that the animals have perfect information about the relative quality of their prey and the rate at which they will encounter them. In many cases, we assume that information about prey is obtained through learning, and foraging decisions are made that maximize the individual's fitness. But details of the learning mechanisms can intrude on this rationale if decisions are dictated more by the state of the animal at the time of learning than at the time of choice.

Humans often make irrational choices in the consumer market that are governed by concerns of past investment rather than future returns. One example of such choice recalls the arguments for continued investment in the Concorde jet and the Vietnam War due to the dollars and lives already invested in the past rather than the probability of success in the future, what Richard Dawkins cleverly called the Concorde Fallacy. A different cause of irrational choice occurs when the value of an item is determined when the individual first encounters that item and the value is not updated in subsequent encounters. A starving animal might place high value on any food item it encounters while near death. It would be irrational, however, to have a preference for that food item later when more valuable food items are abundant. This is referred to as the sunk-costs fallacy ("throwing good money after bad") or state-dependent valuation. This differs from the Concorde Fallacy in that the perceived value of the item is irrational, rather than the strategy for continued investment.

In many animals engaging in many behaviors, there is a decelerating function of value (e.g., fitness) with objective payoff. For example, a prey item of given caloric value will have a greater effect on an animal's prospects for survival when the animal is near starvation than when it is satiated. Thus, as shown in Figure 4.12, if a forager receives a prey item with the same magnitude of benefits (M_L and M_H) when it is in one of two different states, low reserves (L) and high reserves (H), the item encountered when the forager's needs were greater will yield the higher value gains ($V_L > V_H$). The question is do foragers evaluate prey by remembering the value gains, the payoff magnitude, or their nutritional state at the time they learned the prey item?

Locusts are ideal subjects for these studies as they are able to learn their prey, which in this case is plants. In a study by Spencer Behmer, Alex Kacelnik, and colleagues, each locust was conditioned to a wheat seed that was associated with either a lemon or a peppermint taste. These tastes added no caloric value to the food item; the value of each food item was identical. Each locust was conditioned to one type of food when it was in a state of low reserves and the other food when it had high reserves. The food type and the animal's state were balanced across subjects. Once conditioned, the animal was given a choice of the two food items when it was in the two states, low and high reserves. If the locust's decision was based on the payoff magnitude, it should choose randomly between the two food items. If the decision was based on the animal's state when it experienced the food item, then the animal should prefer one item when it was tested with low reserves and the other food item when it was tested with high reserves. Finally, if the foraging decision was based on the animal's "remembering" the value gains, then there should be a preference for whatever food was encountered when the animal had low reserves. The results of these experiments show clearly that the memory of value gains and not payoff magnitude are what determine the forager's choice in these studies (Fig. 4.13).

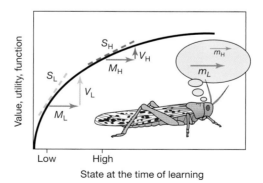

Figure 4.12. A hypothetical model of how state dependence can influence later preferences for food items. The subjects are in one of two states, as shown on the x-axis, low (L) or high (H). The food sources encountered in each condition cause the same magnitude of effect on the animal, M_L, M_H. The values of the effect in each condition are V_L and V_H, and the marginal rates of return in each condition, the slopes of the tangents, are S_L and S_H.

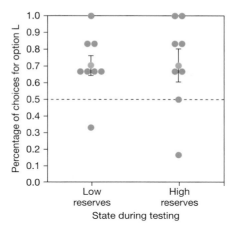

Figure 4.13. Percentage of choices for option L exhibited by each subject under the two different states (low and high reserves) during testing. The figure shows individual data points (red circles) and means (±s.e.m.) (blue circles) with respect to indifference (dashed line). For both states, the percentage of choices for option L was significantly higher than indifference.

One should not assume that state valuation in locusts and humans results from the same mechanisms. There are at least two ways that such decision biases might arise. Remembered value is when the animal has accurate memory of the payoff magnitude of each food item but attaches to each a subjective value dependent on its state at the time. Perception distortion occurs when the animal's perception of the stimulus is influenced by its state. In these animals, for example, nutrient levels in the hemolymph, the fluid that fills the hemocoel, decrease with time since last feeding. As a consequence, the locust's taste receptors become more sensitive to the depleted nutrients. Thus, a hungry locust receives greater sensory feedback when it tastes food than does a satiated locust. Perception distortion might be a more likely mechanism explaining state valuation in insects, whereas remembered value might be more important in humans and some other vertebrates.

Studies that draw attention to mechanisms of learning do not obviate the importance of optimal foraging models. Instead, in this research endeavor, we have another example in which two of Tinbergen's questions—in this case, adaptive significance and acquisition of behavior—need to be integrated to understand why the animal behaves as it does.

Assumptions about Evolution

Adaptations result from the evolution of traits that enhance an individual's fitness. In most studies of adaptation, fitness is usually not measured directly because of logistical difficulties. Instead, some proxy for fitness, a parameter that should be predictive

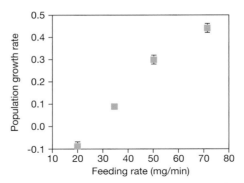

Figure 4.14. The population growth rate in four populations of zebra finches that differed in their feeding rates according to different levels of empty seed hulls in the food. The population growth rate is calculated from age-specific reproduction and mortality.

of fitness, is measured in its place. In optimal foraging theory the proxy for fitness is the relative costs and benefits of foraging decisions. The underlying assumption is that maximizing the rate of energy gain contributes positively to fitness.

This most basic assumption of optimal foraging theory was tested by William C. Lemon with zebra finches. These birds eat seeds. If the seeds are without the hull, the birds eat them readily. Otherwise, they search through the hulls for the seeds. Four populations of birds were all fed the same amount of seeds, but the rate of energy gain was varied. One population of birds was fed 200 g of seeds daily. The other three populations were also fed 200 g of seeds, but the seeds were mixed with 400, 600, or 800 g of empty seed hulls. Thus, birds in the latter three populations had to search for seeds among the hulls. The total amount of food obtained by the populations (E) did not vary, but there were significant differences in the feeding rates due to increased costs of searching (T) related to increased handling time (H). Did this variation in feeding rate influence fitness?

A simple measure of fitness is the rate at which a population increases, because this combines the two critical components of fitness—survivorship and reproduction. In this study, there was a strong relationship between feeding rate and population fitness (Fig. 4.14). The higher the feeding rate, the greater was the increase in population size. At least for this system, the most basic of assumptions about optimal foraging are supported. An optimal forager, one that maximizes E/T, has higher fitness.

The experiments with zebra finches show that the foraging decisions should be under selection because the foraging environment influences the birds' fitness. But selection on a behavior only leads to genetic evolution if there are genes underlying the behavior. It is unusual, however, to know anything about the genetic variation that underlies complex behaviors such as foraging. One exception is in the fruit fly *Drosophila*. These animals are the premiere model for behavioral genetics, and in this system there are genes that have major effects on foraging behavior. Maria Sokolowski and her colleagues have shown that there is a polymorphism in foraging in which "rover" larvae travel widely among food patches in a yeast medium compared to "sitter" larvae. As adults, rovers also move farther from food sources than do sitters. These behaviors result from variation at a single gene with major effects, the foraging, or *for*, gene. Flies with the *for*^{R} allele are rovers, whereas those homozygous for the *for*^{S} allele are sitters. Density-dependent selection influences the relative frequencies of the two alleles. Rover alleles are favored at high-density populations, presumably so flies can forage widely to find food in environments near carrying capacity, whereas sitter alleles are favored at low-density populations in which there is presumably less competition for food. There is substantial information about the mechanisms of the *for* gene, but in the context of optimal foraging, these studies show that there can be genes underlying some very basic aspects of the behavioral repertoire of foraging, and they are favored by selection and evolve in response to selection differently in different environments.

CONCLUSIONS

Finding food is one of the most fundamental ecological decisions made by an animal. This task recruits various aspects of the animal's sensory and cognitive ecology in not only assessing the food but, for some animals, also in remembering where they stored it or how they felt when they first experienced it. We have also seen that animals are able to make very precise decisions that maximize their energy intake and their reproductive fitness. Foraging behavior is one of many behaviors whose intricacies will only yield to an integrative analysis.

BIBLIOGRAPHY

Adamo SA. 2002. Modulating the modulators: Parasites, neuromodulators, and host behavior change. *Brain Behav Evol* **60:** 370–377.

Craig W. 1917. Appetites and aversions as constituents of instincts. *Proc Natl Acad Sci* **3:** 685–688.

Dawkins R, Carlisle TR. 1976. Parental investment, mate desertion and a fallacy. *Nature* **262:** 131–133.

Day LB, Crews D, Wilczynski W. 1999. Spatial and reversal learning in congeneric lizards with different foraging strategies. *Anim Behav* **57:** 395–407.

Emlen JM. 1966. The role of time and energy in food preference. *Am Nat* **100:** 611–617.

Ewert J-P. 1974. The neural basis of visually guided behavior. *Sci Amer* **230:** 34–42.

Ewert J-P. 1997. Neural correlates of key stimulus and releasing mechanism: A case study and two concepts. *Trends Neurosci* **20:** 332–339.

Galef BG Jr, Whiskin EE. 2000. Social exploitation of intermittently available foods and the social reinstatement of food preference. *Anim Behav* **60:** 611–615.

Geffeney S, Brodie ED Jr, Ruben PC, Brodie ED III. 2002. Mechanisms of adaptation in a predator–prey arms race: TTX-resistant sodium channels. *Science* **297:** 1336–1339.

Geffeney SL, Fujimoto E, Brodie ED III, Brodie ED Jr, Ruben PC. 2005. Evolutionary diversification of TTX-resistant sodium channels in a predator–prey interaction. *Nature* **434:** 759–763.

Giraldeau L-A. 2005. The function of behavior. In *The behavior of animals: Mechanisms, function, and evolution* (ed. Bolhuis JJ, Giraldeau L-A), pp. 199–225. Blackwell, Oxford.

Keen-Rinehart E, Bartness TJ. 2004. Peripheral ghrelin injections stimulate food intake, foraging, and food hoarding in Siberian hamsters. *Am J Physiol Regul Integr Comp Physiol* **288:** R716–R722.

Keen-Rinehart E, Bartness TJ. 2007. NPY Y1 receptor is involved in ghrelin- and fasting-induced increases in foraging, food hoarding, and food intake. *Am J Physiol Regul Integr Comp Physiol* **292:** R1728–R1737.

Knudsen EI, Brainard MS. 1991. Visual instruction of the neural map of auditory space in the developing optic tectum. *Science* **253:** 85–87.

Knudsen EI, Konishi M. 1979. Mechanisms of sound localization in the barn owl (*Tyto alba*). *J Comp Physiol A* **133:** 13–21.

Krebs JR, Erichsen JT, Webber MI, Charnow EL. 1977. Optimal prey selection in the great tit (*Parus major*). *Anim Behav* **25:** 30–38.

Krochmal AR, Bakken GS, LaDuc TJ. 2004. Heat in evolution's kitchen: Evolutionary perspectives on the functions and origin of the facial pit of pit vipers (Viperidae: Crotalinae). *J Exp Biol* **207:** 4231–4238.

LaDage LD, Roth TC II, Fox RA, Pravosudov VV. 2009. Flexible cue use in food-caching birds. *Anim Cog* **12:** 419–426.

Lemon WC. 1991. Fitness consequences of foraging behaviour in the zebra finch. *Nature* **352:** 153–155.

Lettvin JY, Maturana HR, McCulloch WS, Pitts WH. 1959. What the frog's eye tells the frog's brain. *Proc IRE* **47:** 1940–1951.

MacArthur RH, Pianka ER. 1966. On the optimal use of a patchy environment. *Am Nat* **100:** 603–609.

Olsen JF, Knudsen EI, Esterly SD. 1989. Neural maps of interaural time and intensity differences in the optic tectum of the barn owl. *J Neurosc* **9:** 2591–2605.

Pompilio L, Kacelnik A, Behmer ST. 2006. State-dependent learned valuation drives choice in an invertebrate. *Science* **311:** 1613–1615.

Sherry DF, Vaccarino AL, Buckenham K, Herz RS. 1989. The hippocampal complex in food-storing birds. *Brain Behav Evol* **34:** 308–317.

Shettleworth SJ. 2003. Memory and hippocampal specialization in food-storing birds: Challenges for research on comparative cognition. *Brain Behav Evol* **62:** 108–116.

Sichert AB, Friedel P, van Hemmen JL. 2006. Snake's perspective on heat: Reconstruction of input using an imperfect detection system. *Phys Rev Lett* **97:** 068105-1–068105-4.

Sokolowski MB, Pereira H, Hughes K. 1997. Evolution of foraging behavior in *Drosophila* by density-dependent selection. *Proc Natl Acad Sci* **94:** 7373–7377.

Migration and Orientation

I N 1976, RONALD REAGAN FAMOUSLY TOLD the American people who were not satisfied with the resources provided by their state that "You can vote with your feet in this country. If a state is mismanaged, you can move elsewhere!" Most people did not leave their state, but they could have; they had the capacity to migrate if they had the desire. In the introduction to this book, we asked you to imagine an animal that does not behave. Now we ask you to imagine an animal that does not move and the consequences of its immobility. Sessile organisms abound across the planet, but they cannot move across it. They are tied to one place with no ability to escape the vagaries of their local environment. This chapter considers the converse—how animals move across space, how they find their way while they do so, and the internal mechanisms motivating and guiding such behavior.

MIGRATION

Animals move all the time and at different scales. Consider an animal foraging for food, as we discussed in the previous chapter, moving across its environment in search of the elusive profitable prey. We can consider its movement at different scales: step-by-step, the path it takes until it finds its prey, and the movement from a profitable hunting ground in the temperate zone in the summer to a profitable hunting ground in the tropics in the winter. In this chapter, we are concerned with movements at the grander scale—the seasonal movement that aids in the animal's survival.

Migration is not a *trait*, but an *attribute* that consists of a set of integrated traits. It involves (1) locomotory activity that is persistent and undistracted; (2) relocation on a greater scale and for a longer time than daily activities; (3) relocation that is usually between regions in which environmental conditions are alternately favorable or unfavorable; and (4) distribution within a spatially extended population.

The world around us is not a stable place. It changes on scales of days, seasons, and geological epochs. Sometimes patience trumps migration; if one prefers the day to the night, it might be better to just wait out the darkness than to circumnavigate the planet to remain in the sun's glow. But in response to other changes within an animal's lifetime, such as those of the seasons, individuals can indeed "vote with their feet" (or their wings or their fins) and migrate to more benign conditions. Animals will go to great lengths to do this, though few as spectacularly as the Arctic tern (Fig. 5.1). This bird has a circumpolar distribution. Its breeding grounds are in the Arctic and sub-Arctic regions. It experiences two summers per year as it migrates to areas around Antarctica for respite from the harsh Arctic winters. Most of the inferences about this bird's migratory habitats came from banding data. For example, an unfledged chick

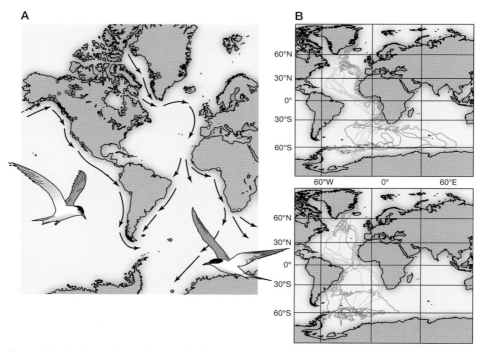

A

B

Figure 5.1. (A) Data obtained primarily from studies of Arctic terns show that these birds migrate each year from as far north as the Arctic to as far south as the Antarctic, a distance of >20,000 km. Some individuals fly (nonstop) across open ocean for >3000 km. (B) Tracking data show details of the migration patterns of Arctic terns in Greenland and Iceland. These figures show interpolated geolocation tracks of 11 Arctic terns tracked from breeding colonies in Greenland (n = 10 birds) and Iceland (n = 1 bird). (Green) Autumn (postbreeding) migration (August–November); (red) winter range (December–March); (yellow) spring (return) migration (April–May). Two southbound migration routes were adopted in the South Atlantic, either (top) West African coast (n = 7 birds) or (bottom) Brazilian coast. Dotted lines link locations during the equinoxes.

that was banded in the United Kingdom was recovered in Australia just 3 months later, a distance of 22,000 km. Given that these medium-sized birds are expected to live for more than 30 years, it is estimated that they travel 2.5 million kilometers in their lifetimes.

Recent advances in tracking devices are now providing insights into long-distance movements of many animals, and a recent study of Arctic tern migrations confirmed their biannual Herculean efforts to skirt between the planet's poles, with some of the 11 tracked birds traveling 80,000 km/yr (Fig. 5.1). These data also showed that although birds from the same colony returned northward on a sigmoidal trajectory that traversed from the east to the west Atlantic, there was variation in the southward migration route. Some southern-bound terns traveled a route hugging the coast of Africa, whereas others used a flyway that bordered along the eastern coast of South America. This polymorphism in migration routes within the same population was not previously known and could provide the raw material for further evolution of migration as we will discuss below.

The Evolution of Migration

There are a number of reasons why animals migrate. The two most important ones are moving to avoid unfavorable climatic conditions and their negative consequences, such as reduced food supply and reduced survivorship during temperate zone winters, and to avoid competition among species for resources. In 1968 George Cox explored these reasons to understand why some birds migrate whereas others do not. In terms of seasonal fluctuations, consider a partial migrant, one that moves within a continuous distribution of its range. Increased seasonal variation might cause the northern and the southern range to constrict (Fig. 5.2). As this occurs, selection might favor an increased migratory segment for the species. Continued change may eliminate the region of year-round favorable conditions completely, and seasonal migration would then occur between seasonally favorable but now disjunct habitats, often one in the temperate zone and one in the tropics. These types of seasonal variation have characterized temperate zones throughout the Cenozoic Era and are thought to have been important in the evolution of migration in numerous bird species.

Another factor that can cause animals to move in space is competition between species. Cox argued that there is a trade-off between time allocation and competition. In the tropics, there are more species, and thus competition among bird species for resources is more intense. Although competition is more relaxed in the temperate zone, the time available for breeding is reduced and usually coincides with the flush of food resources available in the early spring. The result is that clutch sizes of birds are larger in the temperate zone than in the tropics, but survivorship to adulthood is higher in the tropics than in the temperate zone. The conflicting costs and benefits

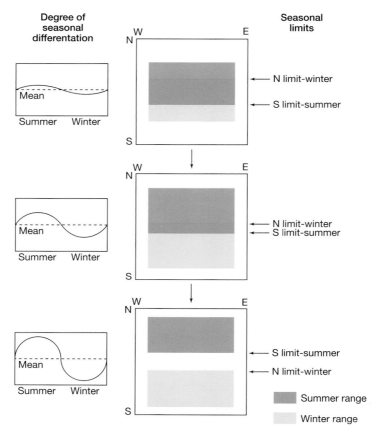

Figure 5.2. How climatic fluctuations can lead to the evolution of migration between two disjunct habitats for a species that is initially a partial migrant. Climate variation causes the favorable habitat to first constrict and then to become disjunct and eventually favor full migration between the disjunct habitats.

result in multiple solutions to the same problem. Some species might respond to this selection by becoming specialized to either the intense competition in the tropics or the harsh conditions in the temperate zone winters, whereas other species migrate between the two, not being as well adapted to either but reaping the benefits of both.

Although Cox modeled the migration phenomenon with a concern for bird ecology, the general arguments apply broadly. The movement between favorable and unfavorable sites also need not be mediated by seasonal variation. Most amphibians, for example, are terrestrial as adults but return to aquatic breeding sites to deposit their eggs. Sea turtles, on the other hand, are aquatic but migrate hundreds of kilometers to deposit their eggs on land. In our own species, many of the more recent mass migrations have been motivated by less favorable economic conditions at home compared to targeted destinations combined with less competition for those resources.

Genetic Bases of Migration and Some Consequences in Evolution and Ecology

Birds have also been a key group in understanding the degree to which there is a genetic basis to migration. Before migration, birds show physiological responses, such as fat deposition, as well as behavioral ones; the behavioral response is called *Zugunruhe*, or "migratory restlessness." An early indication of a genetic basis to migratory behavior is that captive populations of migratory species show that same restlessness at the same time as their conspecifics do in the wild.

Not only are components of migratory behavior thought to have a genetic basis, but also phenotypic variation among individuals is thought to result from genetic variation. In a number of bird species, there are both resident and migratory individuals. In selection experiments with blackcaps (Fig. 5.3), migrant individuals were mated to migrants and nonmigrants to nonmigrants. The migrant line was fixed after three generations, at which point all offspring were migrants. The nonmigrant line was composed of all nonmigrants after six generations. The amount of migratory restlessness within each line is correlated with the probability of being a migrant or a resident.

A review of the quantitative genetics of migratory behavior by Francisco Pulido and Peter Berthold showed that a substantial genetic contribution to migratory behavior in birds is the rule. The average heritability (h^2) of migratory timing in birds is 0.34, and $h^2=0.41$ for the average amount of migratory activity. The mean of various components of the heritability of migratory behavior, including these two, is 0.37.

As we discussed above, climate fluctuations have been hypothesized to play a historical role in the evolution of bird migration. Contemporary climate change due to global warming is also influencing the movement patterns of many animals, as evidenced by changes in the species' ranges. Components of migratory behavior can rapidly evolve a genetic response to rapid climate change. This has been shown with studies of shifts in migration sites in blackcaps.

Blackcaps (*Sylvia atricapilla*) in central European populations traditionally winter to the southwest of their populations in the Mediterranean region (Fig. 5.3). Although there are resident breeding populations of blackcaps in the British Isles, in the past 40 years the number of birds wintering in Britain and Ireland, 1500 km to the northwest of the traditional wintering grounds, has been increasing steadily because of warming of the climate there. This is the result of the evolution of a new migratory route within the blackcaps of central Europe. Orientation bearing in birds is often tested in what has been called an Emlen funnel (a method discussed in more detail below). Birds are put in the bottom of an inverted funnel, the inside of which is lined with paper. The bottom of the funnel has a pool of ink, which leaves marks on the

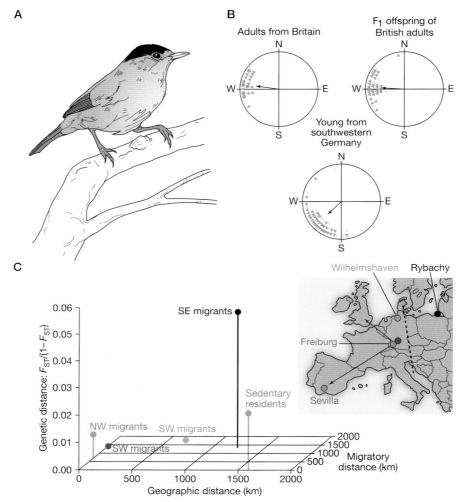

Figure 5.3. (A) A Blackcap, a well-studied migratory species in Europe. (B) The orientation bearings of migrant birds from a population in Britain, their offspring, and young birds from a population in Germany. The northwest bearing of the two British samples is significantly different from the southwest bearing of the German population. (C) In the population genetic analysis of blackcaps, the genetic and geographic distances are illustrated relative to the southwestern (SW)-migrating blackcaps from southern Germany (purple dot). The respective migratory distances are denoted on the z-axis. Each population is characterized by a given color. The map of Europe shows the breeding grounds of the sampled populations and the two pronounced migratory divides in the blackcap. The dotted line corresponds to the migratory divide separating southeastern (SE)- from southwestern-migrating populations. The arrows indicate the distinct migratory orientations in the recently established Central European migratory divide.

paper, leaving an imprint of the direction in which the bird hopped. When migration season commences, the birds hop in the direction in which they would migrate. Berthold and his colleagues tested the bearing directions of adult birds that migrate from Britain, the offspring of those migrant birds, and young blackcaps from Germany (Fig. 5.3). The two samples from the British populations showed bearings to the north-west, whereas the German birds showed a bearing to the southwest, significantly different from that of the British birds. Also, the bearings of the migrant British birds were significantly different from those of the breeding populations of British birds, which show a migration heading to the south. The bearing of the migrant British birds suggests that their home site is somewhere between Belgium and central Germany, an area where banding studies have shown that there are northwest migrants. Thus, there is a genetic basis to the difference in migratory routes of black-caps from central Europe. The first northwest migrants were detected from banding returns in 1961; 30 years later, these migrants constituted ~10% of breeding populations in Germany and Austria. The evolution of this novel migratory route that established a new wintering ground in Britain involves a 50% decrease in migra-tory distance and a shift in direction of 70%. This appears to be evidence of a rapid behavioral response to changing climate conditions that make the British Isles a more favorable place for blackcaps to wait out the winter.

This divergence in genetically based migratory routes could cause some prob-lems. Remember, the blackcaps breed together in central Europe but then separate along their two migration routes to different wintering grounds. What would happen if birds in the same population but with different migratory routes bred? One possi-bility is that some offspring would show the northwest bearing, whereas others would head southwest for the winter. That would not present a problem. But this is not the case. Hybridization leads to a migratory bearing in an intermediate direction, a bear-ing that would send a bird out to sea, southwest of the British Isles, on a trajectory to certain death.

This problem of "hybrid" blackcaps with "hybrid" routes does not occur. It appears that something even more unusual is in the process of unfolding—sympatric speciation. The dogma in evolutionary biology is that speciation most often takes place between populations that are allopatric (i.e., separated geographically). Sympatric speciation, in which a single population diverges into two or more new and distinct species occupying the same geographic range, is usually assumed to be an uncommon event. If so, a rare event seems to be taking place in the blackcaps (Fig. 5.3). Gregor Rolshausen and his colleagues recently conducted a population genetic analysis of five populations of blackcaps: southwest and northwest migrating populations from southern Germany; a southwest migrating population from northern Germany; a sedentary population from Spain; and a southeast migrating population from Russia. The results show greater genetic divergence in neutral microsatellite markers between the two sympatric (southern Germany) migrating morphs, than between either of them and the southeast migrating population 800 km away in

Europe (Fig. 5.3). Thus, the sympatric populations seem to be evolving, sympatrically, into different species distinguished mostly by their winter migration routes.

This genetic divergence appears to result from assortative mating in sympatry. Females do not exhibit active mate preferences based on migratory type; instead, there is temporal asynchrony of arrival at the breeding site between the two migratory types. Thus, individuals at the same locations in central Europe are more likely to breed at the same time as other birds from the same wintering site. The assortative mating and resulting genetic divergence in neutral markers is accompanied by what seems to be adaptive phenotypic evolution. Relative to southwest migrating birds from the same population, northwest migrating individuals have shorter, broader wings, perhaps because of the shorter migration distance; more narrow beaks, perhaps because of diet shifts on the wintering grounds; and browner plumage, for unknown reasons. Thus, the evolution of a novel migratory route has important effects not only on correlated phenotypic traits but also on the origin of species.

We have just seen that in some birds there is variation in the migratory routes taken by different individuals in the same population. There also can be more categorical variation among individuals in whether or not they migrate. This is especially true in insects, where a species can contain a winged form that migrates and a wingless form that does not (Fig. 5.4). In almost all insects that are holometabolous (have distinct larval, pupal, and adult stages [e.g., flies, beetles, moths, and butterflies]) with a winged–wingless dimorphism, the dimorphism is under a simple Mendelian one-gene, two-allele control in which reduced wings are dominant. All of these insects, and almost all insects in general, are in the subclass Pterygota; thus, the presence of wings is the ancestral condition, and loss of wings is the derived state. The loss of wings is also accompanied by the loss of flight musculature. Unlike birds, in which individuals must be banded and recaptured to determine their migratory status, researchers need only to examine their wing structure in these dimorphic insects to determine their migratory status. This advantage combined with the simple genetic system allows some insights into the population genetics of migration not tractable with other systems.

The genetic system controlling wing dimorphism offers an advantage in the face of selection against long-winged migrants because a dominant allele for wing reduction can spread more quickly than a recessive one. The recessive alleles for long wings will be masked by the dominant alleles for short wings in heterozygous individuals. Of course, they still risk loss due to genetic drift. The population genetics of this form of migration was explored in a classical study by V.W. Stein in 1977.

Stein studied four species of weevils. The genetic control of the winged–wingless dimorphism was demonstrated in one of the species, and it was assumed to occur in the other three. Stein followed the number of flightless individuals of each of four species of weevils in newly colonized patches in a recently seeded agricultural field. Not surprisingly, long-winged individuals established the patches first. Short-winged individuals eventually found their way to the patch. In succeeding generations there was a rapid increase in the frequency of short-winged individuals (Fig. 5.4), caused in part by the loss of some long-winged individuals to emigration, but predominantly

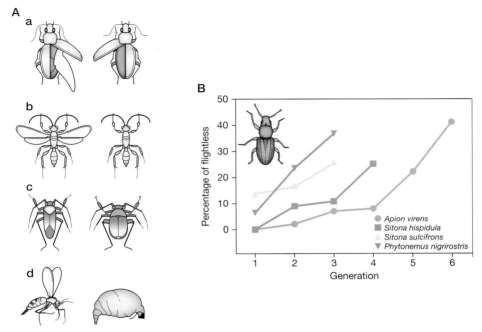

Figure 5.4. (A) Examples of winged (*left*)–wingless (*right*) dimorphisms in (a) the carabid *Pterostichus antracinus*, (b) the hymenoptera *Gelis corruptor*, (c) the bug *Haltitus chrysolepis*, and (d) the fly *Plastosciara perniciosa*. (B) Increase in the percentage of flightless individuals in four species of weevils in patches in a newly seeded field. Unlike most vertebrates, in insects there is a clear correspondence between polymorphism in morphology and migration behavior.

due to the greater fecundity of short-winged individuals. These individuals were able to invest more in reproduction because they did not invest energetically in migration.

Why aren't long-winged individuals, along with their migratory abilities, lost from the population? Simulation models demonstrate that in a heterogeneous environment, in which short-winged individuals cannot colonize new habitats, an equilibrium frequency of the two alleles controlling the wing dimorphism is reached. Unlike migratory patterns in birds, what we see in these weevils, and in many other insects, is a genetic polymorphism with pioneers that first settle new habitats followed by sedentary settlers that then further populate the new habitats. The heterogeneity of the habitat is combined with the simple genetic control system, ensuring that neither form goes extinct.

Physiological and Neural Preparations for Migration

The example of the Arctic tern's global travel described at the beginning of this chapter highlights the impressive behavior typifying animal migration. Birds fly across oceans and continents. Salmon travel hundreds of kilometers, in part upstream through strong river currents, on their final, fatal adult migration home. The physiological demands of long-distance migration are incredible even when animals have the opportunity to stop periodically and feed. But without dramatic seasonal adjustments in their regulatory physiology, birds would not be capable of the long-distance migrations they undertake.

The use of energy reserves by vertebrates generally follows a progression from using stored carbohydrates (glycogen, which is depleted on a timescale of a few hours), to using stored lipids (from body fat, which can take days to deplete), then to proteins via the breakdown of muscle. Even relatively short migrations of several hundred kilometers would deplete glycogen stores. Passerine birds in general have little glycogen stores with which to work. Reaching the stage of protein breakdown is dangerous if it impacts flight muscles. That leaves body fat as the preferred storage energy source.

As the start of the migratory season approaches, passerine birds enter a period of hyperphagia along with their general *Zugunruhe* (the migratory restlessness discussed above), in which they can increase their daily calorie intake by >50%. The increase in eating may coincide with a change in food preferences. Insectivorous birds often switch to eating fruit prior to migration. This may relate to seasonal availability; however, fruit is more efficient digestively than insects with all their indigestible chitin, and, if abundant, fruit requires less energy expenditure to acquire. As a result, largely driven by the increased eating, premigratory birds increase body weight and, more spectacularly, body fat. Whereas passerine birds normally average 5% body fat, premigratory birds average 24% body fat, and some transoceanic migrators (which may fly up to 3 days nonstop) can reach 70% body fat. The hyperphagia is supported by, and likely is largely responsible for, an increase in the size of the digestive tract, which can enlarge in surface area by a factor of 3. This gut enlargement allows the bird to maintain efficient extraction of nutrients from the high volume of ingested food.

There is little evidence for changes in digestive enzymes to enhance nutrient recovery unless there is a dietary switch, which should result in reactive changes in digestive enzymes, especially when shifting between high-protein/fat and high-carbohydrate diets. Very shortly before migration begins, long-distance migrators switch to hypophagia, or fasting. The lack of eating results in regression of the gut and other digestive organs such as the gizzard and liver, which reduces body weight, reduces the amount of a highly metabolically active tissue, and provides protein-based energy while sparing muscle tissue used for flight. Migrating birds are often slow to replace the body weight they lost during the migratory travel when they arrive at their destination, even when given access to unlimited food. The regressed gut accounts for this. It takes at least several days for the gut to return to its premigratory size.

Changes in other tissues accompany the digestive changes. In some birds, flight and cardiac muscles hypertrophy in preparation for flight. Molting, or the replacement of flight feathers, is commonly timed to occur just before migration. When additional environmental challenges must be faced, other physiological changes occur. Salmon and sea trout are born in freshwater and live as juvenile parr, then transition into smolt, which migrate into oceanic saltwater. They then go back to their freshwater spawning grounds as adults. Freshwater and saltwater fish face very different challenges to maintaining their ionic and osmoregulatory balance. In freshwater, fish must retain sodium and chloride ions and excrete large amounts of water as urine, whereas in saltwater, fish must excrete both ions against a strong concentration gradient and retain water. These require opposite activities by ion pumps in gill cells, in kidney and intestinal function, and in the endocrine regulation of the entire system, all of which are reorganized wholesale during the smolt transition, then back again during the transition from smolt to sexually mature adults.

Emergent Properties and the Origins of a Mass Migration

Mass migrations of animals have long been viewed as among the truly spectacular events in natural history. In his 1831 account of the now extinct passenger pigeon, John Audubon notes the incredible size of these migrations: "The air was literally filled with Pigeons; the light of noon-day was obscured as by an eclipse, the dung fell in spots, not unlike melting flakes of snow; and the continued buzz of wings had a tendency to lull my senses to repose." The mass migration of insects can be plague-like and literally of Biblical proportions:

> And the locusts went up over all the land of Egypt, and rested in all the coasts of Egypt: very grievous were they. . . . For they covered the face of the whole earth, so that the land was darkened; and they did eat every herb of the land, and all the fruit of the trees which the hail had left: and there remained not any green thing in the trees, or in the herbs of the field, through all the land of Egypt.   —Exodus 10:14–15

In addition to their size, mass migrations are often notable for their resemblance to a superorganism with the movements of individuals so well coordinated they appear to be organized at the group level. In many cases, however, the complex group-level patterns appear to be the result of self-organization, an emergent property of the behavior of individuals. Thus the perceived goal of the group might have little to do with the actual goal of the individual. In no system is this better understood than in the flightless Mormon crickets and their close relative the desert locust.

Both the flightless Mormon crickets and the flightless juvenile desert locusts form migratory bands that can be >10 km long, several kilometers wide, and with millions of individuals that march lockstep at a rate of ~2 km each day. Greg Sword and his colleagues showed that the bands seem to form in part from a selfish herd advantage, as there is protection in numbers. Mormon crickets that were translocated from the band were more likely to suffer predation. Thus forces external to the group are responsible for the clumping in space—join the band and you are likely to live longer.

It is forces within the band, however, that keep it marching. Animals that lag behind get eaten. Crickets in migratory bands were shown to be salt and protein deprived. The most immediate sources of both are near at hand; crickets are delectable packets of protein and salt, and they are notoriously cannibalistic. Any individual that lags behind quickly becomes a food source for its hungry neighbors. The greater is an individual's lack of protein, the more it moves. Feeding protein and salt to crickets makes it less likely that they will cannibalize their neighbors. Thus, once some animals start to move in response to lack of protein, others must move as well to keep from being eaten.

Marching bands of juvenile desert locusts are similar to those of the crickets, including the potential role of cannibalism in group formation. Its importance was demonstrated clearly with ablation experiments. Researchers severed the ventral nerve cord immediately posterior to the metathoracic ganglion. This ganglion is involved in transmitting information from sensory receptors on the abdomen, including the cerci at its caudal end, to the brain. The denervated animals, therefore, could not detect the approach of neighbors from behind. These ablation experiments showed that attention to the rear guard, the potential cannibals, was critical both for group function and individual survivorship. When such attention was deprived, it decreased the individual's probability to start moving, dramatically reduced the mean proportion of moving individuals in the group, and significantly increased cannibalism.

Thus, we can see that the large migratory bands of these crickets and locusts result from an individual's search for protein and salt and its need to avoid becoming the source of that protein and salt for its neighbors. Thus, the behavior of the group is an emergent property of the behavior of the individuals within it. This argument does not, however, explain two important characteristics of such groups. One is the juvenile locust's transition from a solitary individual to a group member, and the other is the precise alignment and coordinated movement within the group.

The transition from a benign solitary stage to an aggressive gregarious stage is density dependent, with formation of the marching band and precise alignment of movement among neighbors happening only at high densities. In laboratory experiments, coordinated marching behavior did not occur until densities of more than 70 locusts per square meter were reached. The most obvious interpretation is that at high densities individuals behave differently and in a manner that results in group cohesion. But there appears to be a more fundamental explanation.

Studies of self-organizing systems attempt to understand how interactions at the individual level can result in emergent patterns at the group level. These density-dependent effects on group migration in locusts are strikingly similar to some recent attempts to predict group behavior as an emergent property of individuals behaving as self-propelled particles (SPPs). The idea is that individuals adjust the speed and direction of their movement in response to their neighbors. Once a critical density is reached, there then is a rapid transition from disordered movement of individuals to highly aligned collective movement. This is despite the fact that the behavior of individuals does not change with density; they do not at some point decide to behave as a member of a group, but instead continue to behave egocentrically.

An SPP model by Jerome Buhl and coworkers incorporated parameters similar to the movement behavior of the locusts and was used to predict how locust density influences three variables: (1) mean alignment, scaling from +1 when individuals are aligned in a counterclockwise manner to −1 when they are aligned in a clockwise manner; (2) how much time individuals spent aligned with neighbors; and (3) how often there was a change between an aligned and an unaligned state. The data generated from the simulations were compared to the experimental results, and the two were consistent for all three variables (Fig. 5.5). The group characteristics of the locusts—increased alignment, time in alignment, and decreased change in alignment—all were more pronounced at higher densities and exhibited a rapid transition rather than a gradual change. Also, both model and data exhibited dynamic instability as changes in direction were rapid and spread spontaneously throughout the group, and once a group is gregarious its behavior does not change. These same behaviors are seen in desert locusts in the field. Critically, the behavior of individuals did not change with density, but the emergent parameters of the group did. As the SPP model predicts, collective group-like behaviors do not result from a change in behavior at the individual level, but merely from increased density.

Environmental and Physiological Triggers of Migration

Migrations in general are a mechanism of resource management, by which an animal avoids a resource-poor region by moving to a resource-rich region. In the example given above, insect mass migrations are triggered by internal signals of resource decline, namely, a nutrient deficit. When the resource changes are seasonal, animals can take advantage of external seasonal cues that predict impending deficits and act accordingly rather than wait until they are in energetic distress to begin their movement. For birds and other seasonal migrators that routinely move back and forth, day length is the primary environmental cue influencing the behavioral and physiological changes preparing an animal for its migration. Birds, and for that matter fish and other long-distance migrators, are not, however, passive responders to environmental signals. Instead, there are clear indications that animals possess an intrinsic, endogenous, circannual rhythm that controls their annual life cycle. Migratory birds held in constant environmental conditions will still molt, express *Zugunruhe*, and fatten at about the appropriate time of year (Fig. 5.6). Environmental cues interact with the endogenous rhythm by either synchronizing it to the external year (endogenous circannual rhythms have routinely been found to be longer than a standard year, and individually variable), or limiting expression of an endogenously controlled annual behavior to a certain environmental condition. Circadian rhythms also work this way: An endogenous biological rhythm is entrained or gated by external environmental cues.

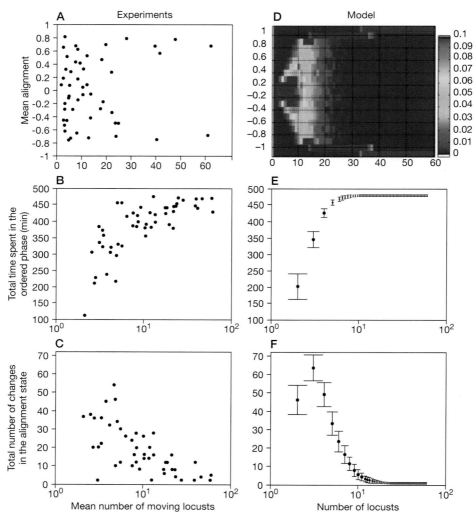

Figure 5.5. The actual alignment of locusts as a function of density (*A–C*) compared to model pre-dictions (*D–F*). The number of moving locusts and their mean alignments are represented for the experiments (*A*) and for simulations (*D*). Each point in *A* represents the mean of one experimental trial. Each of the colored columns in *D* represents the distribution of the outcomes of 1000 simula-tions. For further analysis, alignment was categorized into three states: unaligned (alignment value between −0.3 and 0.3), aligned counterclockwise direction (alignment value 90.3), or aligned clock-wise direction (alignment value G − 0.3). A change in state required that it persisted for 1 min (16 time steps in the simulation) or more. *B* and *E* show the relationship between the average number of moving locusts and the mean total time spent in the aligned state for the locusts and model, respectively, whereas *C* and *F* show the mean number of changes in the alignment state on a semi-log scale for the locusts and model, respectively. Error bars, standard deviation. These results show that changes in group behavior that occur with density are an emergent property and do not result from changes in the behavior of individuals.

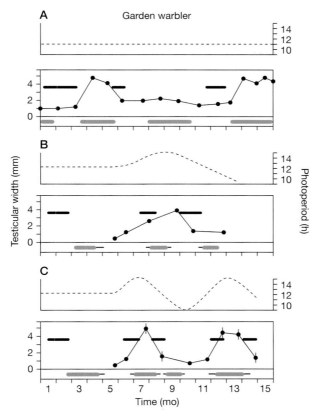

Figure 5.6. Seasonal patterns of testicular change (black line), molt (black bars), and *Zugunruhe* (red bars) in Garden Warblers over 15 mo. Testicular size is indicated on the y-axis; bars simply indicate the presence of molt and *Zugunruhe*. (Dashed line) Change in day length over 15 mo. (A) Under a constant light–dark cycle, seasonal testicular, molt, and migratory restlessness still appear in a roughly circannual pattern. (B) When day length changes on a 12-mo period, the bird's endogenous cycles entrain to the lengthening day. (C) When the day length cycle is changed to a 6-mo period, the bird still entrains to the shortened "year."

Hormonal changes are important in mediating the link between a migratory trigger and the coordinated suite of behavioral, physiological, and anatomical changes that occur. In vertebrates, thyroid hormones, particularly thyroxine, appear to be the most common hormones regulating the changes that prepare and motivate an animal for migration. This is true for both seasonal migrations and for migrations, such as those in salmonid fishes, that are part of a developmental cycle. Gonadal steroid hormones may also be involved when migration is related to the onset of a breeding season. In insects, migrations are also often part of a developmental sequence related to adult dispersal. In invertebrates, juvenile hormone is the most common regulator. It is interesting to note that thyroxine in vertebrates and juvenile hormone in insects are important in other life cycle changes such as metamorphosis, which, like migration, involve significant changes in behavior, anatomy, and physiology.

ORIENTATION AND NAVIGATION MECHANISMS

Navigation for both seasonal migrations and shorter-distance episodic travels requires several components to be successful. If the function of either is to travel from, then back to, a specific location such as a home territory or natal location for reproduction, the animal must have knowledge of characteristic features of that location so that it knows that it has returned to the correct place. When at a distance from the terminal location, it needs to know both where it is in relation to its goal and what direction to travel.

Not all animals move of their own accord, however. Many passively disperse with their environment. Numerous species are thought to reach distant islands by rafting on tops of mats of vegetation carried by ocean currents. Many high-flying insects also use a form of aerial rafting, as they are dispersed by high-altitude air currents that transport them at speeds and distances that far exceed what would be possible by their self-powered flight. In such cases, animals cannot orient, and need not have any reason to do so—they are at the mercy of the environment. Or so it was thought.

Subtle Orientation

Many moths disperse on high-altitude currents to reach migratory destinations hundreds or even thousands of kilometers away. In a study that used radar tracks of mass migration events, researchers showed that in initiating their northward migration in the spring, the moths' headings tended to coincide with the prevailing northerly direction of the wind. In the fall, however, when the moths migrate to the south, there are prevailing easterly winds. In these cases, the moth's headings were significantly different from the prevailing direction of the wind. Their headings corrected for the easterly winds with the result that the moths reached their predetermined site, which would not have been possible if they passively headed off the prevailing wind. The effect of the moths' flight behavior on their aerial rafting was shown with simulation models. The trajectories of inert particles that would travel passively with the wind were compared with the trajectories of one of the species of moths given the actual headings and flight speeds it used. The results showed that the behavior of the moth substantially increased the distance traveled and the accuracy in reaching its goal site (Fig. 5.7). Thus, this example of apparent passive dispersal shows subtle but critical orientation behavior.

These high-flying moths can take control of their flight direction, but how do they know where they are and where they want to go? The researchers suggest that the

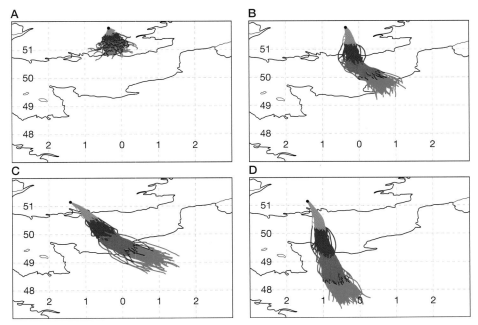

Figure 5.7. Graphs show simulations of 8-h flights (color-coded in 2-h segments) for 100 inert particles (*left, A,C*) and 100 moths (*right, B,D*) released from two radar sites in southern England (Rothamsted, Hertfordshire, *A,B*; Chilbolton, Hampshire, *C,D*). The moth's flight behavior enhances the distances traveled by merely "aerial rafting," and an increase of 159.7 km in *A* versus *B*. Accuracy of flight is also enhanced by the moth's behavior as seen in the *bottom* panel, where the flight heading of the moths (169.0°) compared to the particles (144.8°) brought the moths 24.2° closer to their migratory goal.

animals must have an internal compass, but a compass without a map is of little help. The mechanisms of orienting in these and many animals remain a puzzle, leading to the question, how do animals know where to go?

Olfactory Imprinting

Locations have distinct smells, sounds, sights, and feels, and animals can use any of them as cues. The first three are useful for navigation because they can be detected at a distance, and animals can then use them as navigational beacons. Smells and sounds are better for long distance as they are less disturbed by intervening objects. Of these, the potential persistence of odor cues in their transmission through the environment plus the extraordinary sensitivity most animals have to odors make them a particularly important signal for many cases of animal migrations.

Pacific salmon of various species make a remarkable life history migration that depends on olfactory imprinting. As noted above in reference to the dramatic physiological and sensory changes involved in this migration, fish such as coho and sockeye salmon begin life as larvae, and then as postmetamorphic parr in freshwater lakes and rivers for their first 18 mo. They then transform into smolt, which migrate downstream to enter an ocean habitat, where they live for 3–4 yr, growing into adults. Their ocean migration can take them thousands of kilometers away from the river mouth from which they emerged, but upon reaching sexual maturity, a second transformation begins, bringing them back to their natal freshwater habitat to spawn. A large number of marking and tracking studies have consistently confirmed that 95% of the adult salmon return to spawn in their native streams, finding the exact tributary in which they were born 5 or more years before.

In 1951 Arthur Hasler suggested that fish imprint on the characteristic odors of their home waters early in life and then use olfaction to find their way home. Imprinting is a particular type of learning in which an organism, early in life during a critical period, rapidly and permanently learns the characteristics of some object or stimulus. Imprinting was discovered by Douglas Alexander Spalding, a 19th-century lawyer, philosopher, and biologist who noted the way in which young chickens bonded to, or "imprinted on" an adult based on their early experience. Konrad Lorenz later took up the phenomenon of imprinting, describing how geese would "imprint" on individuals they saw at a critical period shortly after hatching—in one famous demonstration, on Lorenz himself. (We will return to imprinting briefly in Chapter 8, where we discuss social bonding and the role of early experience with odors in mother–infant recognition.) Several experimental studies using artificial odors, many by Hasler and his students over his long career, have provided strong support for the idea that olfactory imprinting is also important in salmon homing.

One very large experiment ultimately used more than 45,000 coho salmon from the same genetic stock in a hatchery that breeds and releases an introduced population of fish into Lake Michigan (Fig. 5.8). One group of coho smolts was exposed to lake water containing morpholine (an artificial chemical not found in streams), one group to lake water with β-phenylethyl alcohol (PEA; another such artificial chemical), and a third to untreated lake water. All were released into Lake Michigan midway between two test streams opening into the lake. One year later during the time of the return migration, a small amount of morpholine was drained through one test stream and PEA through the other. Of the fish recaptured in each stream, 95% of the morpholine-exposed fish were captured in the stream containing morpholine, and 93% of the PEA-exposed fish were captured in the stream containing PEA. Few of the control fish were captured in either stream, presumably because both streams had olfactory characteristics unfamiliar to them.

The parr-to-smolt transformation appears to be the critical period for olfactory imprinting. This is a time when thyroxine levels rise dramatically, which, as you may recall from above, is the thyroid hormone responsible for orchestrating numerous physiological and anatomical changes associated with migration as well as metamorphic transitions. This elevation of thyroxine is also important in triggering olfactory

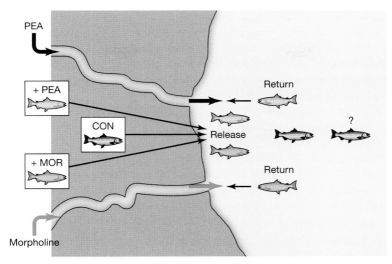

Figure 5.8. Experiment demonstrating olfactory imprinting in migrating salmon. Cohorts of salmon from the same genetic stock are exposed to PEA (β-phenylethyl alcohol), morpholine (MOR), or control (CON) water from Lake Michigan during the parr–smolt transition, then released together midway between two streams draining into Lake Michigan. Recapture during the return migration revealed that PEA- and morpholine-exposed fish entered streams in which those chemicals were released. Few fish exposed to lake water alone entered either stream.

changes critical for salmon imprinting. Allan T. Scholz, a student of Hasler, found that elevating thyroxine levels artificially through injections of thyrotropin-releasing hormone in salmon parr before their normal smolt transformation, then pairing that elevation with morpholine or PEA, caused the fish to imprint on that odor. In later tests, fish swam toward a stream branch containing the artificial order that had been paired with the induced thyroxine surge. Recent work has pinpointed the mechanism of olfactory imprinting to the olfactory receptors themselves. Olfactory receptors turn over with every 10–14 d in all vertebrates and are replaced by new receptors derived from basal cells in the olfactory epithelium. Thyroxine stimulates proliferation and development of basal cells, and such increased proliferation occurs in salmon during the parr-to-smolt transition. Presentation of chemical cues during this time results in increased peripheral olfactory sensitivity to those odors, which, in experimental studies, is evident electrophysiologically for months afterward. Olfactory imprinting, therefore, is a selective change in the sensitivity of the olfactory system to particular odors, mediated by a change in the organization in olfactory receptors.

The utility of using waterborne chemicals for homing in fish such as salmon makes sense ecologically. Fish returning from their ocean feeding period can search river outlets until they detect the particular odorants flowing out of them from the natal lakes. They can then follow the upstream odor trail, and at river branches orient to and follow whichever branch contains the odorant until reaching the source. But this process cannot explain how fish can travel through the open ocean to the correct coastal location where they could find the odor. Similarly, it cannot explain other instances of long-distance migration found, for example, in migrating birds or sea turtles. That process requires a more complicated form of navigation, which will be taken up in the next section.

Navigation by Compass and Maps

True navigation requires an organism to move from its present location to a destination without homing directly toward a (more or less constantly) detectable signal. To achieve such a task, one needs to know two things: basic information about directions (e.g., which way is west?) and knowledge about your location relative to your destination (e.g., am I north or south of home?). For the first, you need a "compass"; for the second, you need a "map."

As discussed above, during the premigratory *Zugunruhe*, birds become agitated and hop in the direction that they would normally migrate. But how do they know which direction that is? Animals use a variety of mechanisms to determine direction. This can be demonstrated by placing them in an Emlen funnel or other circular arena while restricting or manipulating the cues available to them. Such experiments reveal a variety of stimuli that animals use as compasses, from celestial compasses such as the sun, stars, and polarized light patterns, to magnetic compasses using the earth's magnetic field.

Sun and Star Compasses

Many birds are *diurnal* migrators; that is, they fly during the day and rest at night. Other birds such as pigeons that have a strong homing instinct are also diurnally active. For them, the sun can be used as a compass. Diurnal migrators such as starlings (*Sturnus vulgaris*) will orient and attempt to fly in their normal migratory direction even when placed in a test cage where only the sun and sky, but no other visual landmarks, are visible. Shifting the apparent position of the sun via mirrors will shift the birds' orientations. Many small songbirds fly at night and stop to eat and rest during the day. Stephen Emlen's classic paper, for which the Emlen funnel described above was developed, showed that indigo buntings (*Passerina cyanea*) use star patterns as a compass (Fig. 5.9). Emlen used a planetarium projector to present star patterns above the buntings' cage. Shifting the star map shifted the orientation patterns, and random orientations were exhibited by birds exposed to a blank "sky."

Sensing Polarized Light as a Compass

One additional celestial compass is the pattern of polarized light in the sky. Polarization patterns occur when the molecules in the atmosphere scatter light and are most apparent for short-wavelength light (blue to ultraviolet). Light waves consist of

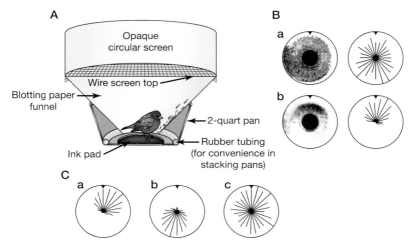

Figure 5.9. (A) Emlen funnel apparatus used to determine the orientation of indigo buntings during migratory restlessness. The bottom contains an ink pad, and the top is an opaque screen on which star patterns are projected. (B) Birds leave ink stains on the sloping sides when hopping toward their preferred direction: (a) random when no stars are projected, (b) northerly prior to spring migration. Ink patterns and vector transcription for each are shown. (C) (a) Resultant vectors in the spring under spring star pattern; (b) under a star pattern rotated 180°; (c) with no stars visible.

electromagnetic fields that oscillate perpendicular to the direction that light is traveling. The orientation of these fields is called the "e-vector." As the light interacts with atmospheric molecules, the e-vectors of some proportion of the light become oriented, or "polarized," in the same direction. The polarization lines form concentric circles around the sun, are more noticeable the farther away they are from the sun in the sky, and are most apparent on the horizon at dawn and dusk. If animals can see those patterns, they can use them like the grid lines of a map superimposed on their vision of the world. Many animals, including birds, fish, crustaceans, and insects, have visual systems that can detect light polarization patterns. Comparing how invertebrates and vertebrates accomplish this (Fig. 5.10) is instructive as an example of the convergent evolution of a behavioral capacity, with the same ultimate function, but with different mechanistic characteristics due to the constraints of very different sensory structures.

Photopigment molecules absorb light best when they are oriented in the same plane as the e-vector. This is a trait shared by both vertebrate and invertebrate photopigments, so in that sense, all animals are potentially able to "see" polarized light. However, if the photopigments in a receptor cell are oriented randomly within it, the receptor simply sums across all pigment activation and comes up with the total light energy in that particular part of the visual field, losing any sensitivity to the direction to which the e-vectors within that light are oriented. Invertebrates and vertebrates solve this problem in very different ways. Invertebrates have compound eyes

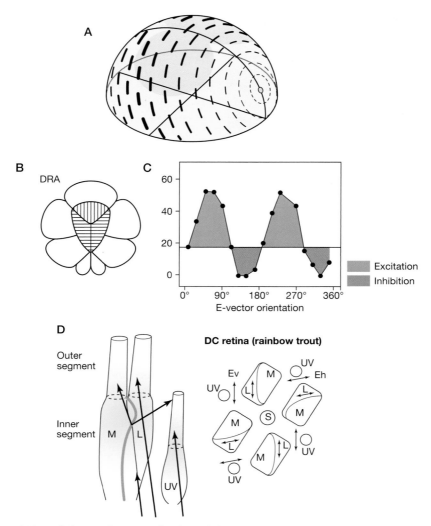

Figure 5.10. Cellular mechanisms of polarized light detection in invertebrates and vertebrates. (A) Polarization light patterns appear in the sky in concentric circles (black dashed lines) around the sun (yellow dot). The darkness of black lines indicates the strength of polarization: The strength increases at progressive distances from the sun. Dark blue circular areas indicate area of the sky that would be seen by the dorsal rim area (DRA) of an insect's compound eye. (B) Organization of photoreceptor cells in an ommatidium in the DRA. Lines indicate the orientation of microvilli containing photopigment. (C) Schematic response pattern of an optic lobe neuron receiving excitatory input from photoreceptors with one microvilli orientation and inhibitory from photoreceptors with the orthogonal microvilli orientation plotted against the e-vector (polarization) orientation of a light stimulus. (D) (Left) Organization of double-cone and ultraviolet receptor in the retinal mosaic of rainbow trout and proposed mechanism of polarization sensitivity. (See facing page for legend.)

composed of ommatidia, each a compartment containing a few photoreceptors, and each photoreceptor has photopigment molecules fixed on microvilli within it. Insects such as crickets and ants have a specialized part of their compound eye, the dorsal rim area (DRA), where each ommatidium contains some receptor cells with microvilli strictly oriented in one direction and some with microvilli oriented orthogonal to their partners. Thus, each receptor cell would be most sensitive to light polarized in a particular direction 90° from that of its neighbor in the ommatidium. Optic lobe neurons in the brain take information from both receptor types in an ommatidium and are excited by one set and inhibited by their perpendicular partners. In this way, the optic lobe neuron is strongly excited when the ommatidium is lined up with the e-vector (the polarization plane) in one direction, strongly inhibited when lined up with light polarized 90° to this plane, and has an intermediate firing rate when lined up at a 45° angle to the e-vector. e-vector orientation (polarization) is thus translated into differing degrees of perceived brightness depending on how well the vector aligns with the photopigment molecules on the microvilli. Different parts of the DRA have ommatidia oriented in different directions so that when scanning the sky this part of the compound eye can determine the polarization patterns, and the insect's orientation relative to it, based on the pattern of strongly excited and inhibited ommatidia.

Although it is clear from behavioral studies that many vertebrates can detect polarization patterns, their mechanism for doing so remains obscure. Vertebrate photoreceptors are unable to detect polarization patterns in the same way as invertebrates owing to their cellular construction. In all vertebrate photoreceptor cells (the rods and cones), photopigment molecules are located on flattened discs stacked in the cells. The photopigment molecules within the discs can rotate and move randomly within it, thereby eliminating the ability of a single receptor cell to detect the orientation of

---

Figure 5.10. (Continued) (M, L) Middle- and long-wavelength sensitive cones; (UV) ultraviolet-sensitive cone. "Outer segment" is the location of the disks containing photopigment molecules; "inner segment" is the cell body of the receptor neuron. Arrows indicate the path of light. The thick black line shows the location and orientation of the partition membrane separating the cone pairs of the double cone. The partition's properties make it a mirror that selectively reflects polarized light of a particular orientation. When light of that particular polarization enters the photoreceptor level, the UV cone is stimulated by the directly received light and by light reflected by the partition, and the L and M double cones receive different amounts of stimulation due to the splitting of the light by the partition membrane into reflected and refracted components. Light that is polarized 90° from this orientation will not be reflected and the UV cone will be stimulated only by the directly received light, whereas the L and M double cones will receive equal light stimulation. In this way, polarization differences result in differences in the amount of light striking the UV cone as well as a different ratio of L and M activation. (Right) Each subarea of the retinal mosaic has double cones and associated UV receptors oriented in different directions. Double arrows indicate the e-vector orientation plane sensitivity in the double cones and associated UV receptor (Ev, vertical plane sensitivity; Eh, horizontal plane sensitivity). Higher-order neurons that can compare activity in different UV cones within the area can determine the direction of polarized light patterns. Short-wavelength cones (S) within the mosaic do not receive any reflected light; as predicted, they show no sensitivity to polarization patterns.

the e-vector since it sums the light activation over the hundreds of discs within it. It may be that vertebrates have evolved multiple solutions to detecting polarized light, but Craig W. Hawryshyn and coworkers have examined one potential mechanism that explains polarization sensitivity by ultraviolet cones in fish. A section of the trout retina has a precisely organized mosaic of photoreceptor cells (cones) in repeating units composed of "double cones" and adjacent single-cone UV photoreceptors. Each double cone consists of a medium-wavelength-sensitive and a long-wavelength-sensitive cone fused together. The cell membrane that serves as the partition between the fused cones has peculiar properties that cause it to act like a "dielectric mirror," that is, a surface constructed of multiple layers whose physical and electrochemical properties cause it to pass and refract a portion of light (up into the discs of photopigments above it) and reflect and concentrate another portion of light of a particular wavelength, and, more importantly here, a particular polarization, away from those discs. The partition membrane is anatomically structured so that its upper portion is tilted relative to the pathway of incoming light in such a way that reflected light is cast onto the UV cone adjacent to it.

Each repeating subunit of the retinal mosaic consists of a set of double cones with their reflective partitions oriented orthogonally to each other. Different subunits have double cones oriented in different directions. In one of the double cones in the subunit, light of one polarization orientation will be reflected onto a UV cone, adding energy to the light it is already receiving and making it see a brighter light. Light polarized 90° relative to it will not be reflected, no additional light will hit the UV receptor, and the light the UV cone sees will be less bright. Light polarized 45° relative to the first polarization will reflect an intermediate level of additional light, and the UV cone will see intermediate brightness. For a double cone in the same mosaic unit with a reflecting surface oriented orthogonally to the first double cone, its UV cone will see dim light, bright light, and intermediate light at the same time. Subsequent neurons receive excitatory input from one UV receptor and inhibitory input from another within the subunit. Thus, strong excitation results from light with one e-vector orientation, strong inhibition results when the vector is rotated 90°, and intermediate excitation results from orientations in between.

Just as in invertebrates, vertebrates transform polarization patterns into perceived brightness patterns (but through differential reflection rather than fixed molecular orientation), construct subunits of orthogonally oriented detectors (but through a mosaic consisting of repeating subareas of receptor cells on the retinal surface rather than cells contained in a discrete ommatidium), and depend on higher-order neurons to compare the outputs from orthogonally sensitive receptor cells to determine the e-vector direction, or polarization of the light hitting the retina.

Earth's Magnetic Field as a Compass

One of the most surprising discoveries in sensory biology is that many animals are sensitive to the earth's magnetic field and can use magnetic information as a compass just as humans use magnetic compasses for directional information. Although there is solid evidence from behavioral experiments, some of which were reviewed in Chapter 1, that many animals from diverse vertebrate and invertebrate taxa can sense a variety of magnetic field features and that they can use this information for navigation, there remains much to be done to understand the neurobiology of magnetic sensing.

Two types of magnetic sensing systems have now received considerable support. The most widely supported is based on the presence of microcrystals of magnetite, a mineral form of ferric oxide, which act as microscopic bar magnets (Fig. 5.11). Magnetite crystals were first found in a variety of bacteria that orient themselves and move along magnetic field lines. They are also now known to be present in a large number of diverse vertebrates and invertebrates that have been demonstrated to use magnetic fields for orientation. In birds, magnetite deposits are found above the beak associated with neurons of the ophthalmic branch of the trigeminal nerve, and axons in that branch fire in response to magnetic field changes around the animal. The cellular mechanism is still unknown but, in principle, could be mediated by magnetite crystals being connected to ion channels in neural membranes so that as the crystals twist to align with a magnetic field line (in the same way the needle on a compass moves to align with magnetic north), ion channels open or close accordingly, activating and changing the electrical state of the cell.

A second mechanism is based on the effect of magnetic fields on cellular chemical processes. This can occur when the process involves electron transfer between pairs of radicals, but for the earth's magnetic field only under the right conditions. It happens, however, that processes associated with photoexcitation of pigments in visual receptors can meet the requirements for these effects. A third potential process, and the earliest advanced, suggests that the electroreceptive system of marine animals like sharks could function as a magnetic detector as well, as movement of a conductive circuit (like an electroreceptor) through a magnetic field induces a current that

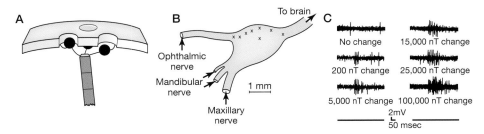

Figure 5.11. (A) Magnetite-based magnetic reception could result from single crystal magnetite attached to ion channels in neural membranes. As the crystal twists to align with the earth's magnetic field, ion channels are pulled open, resulting in a neural response. Sensory neuron activation through mechanical opening of ion channels is also found in the auditory, vestibular, and somatosensory systems. (B) Electrophysiological recordings of magnetic sensitive trigeminal neurons (marked by ×) in the bobolink associated with the beak, where magnetite deposits are found, confirm magnetic sensitivity. (C) Neural firing increases transiently with progressively larger changes in magnetic field strength.

could be detected. Electroreception was discussed in Chapter 3. Currently, this hypothesis for magnetic reception in marine organisms is not well supported.

Circadian Clocks and Compass Calibration

Migrating animals routinely use a variety of compasses to direct them. Birds, for example, often use celestial cues as their first priority, but will then use a magnetic compass if those cues are absent or ambiguous. One additional factor complicating the use of any celestial compass is that the positions of sun, stars, and even polarization patterns change during the day and over the course of a year, and even magnetic information varies geographically because of local anomalies. Animals must have some way to compensate for these vagaries, and, in fact, there is solid evidence that one component of this is by use of the internal circadian clock that all organisms have to regulate daily rhythms. Experiments that artificially shift an animal's circadian rhythm caused birds to misinterpret the sun's position so that they oriented toward the direction based on where the sun would be according to their shifted internal clock, rather than where it actually was on the actual time and day of the experiment. Another strategy is to use a compass cue at a particularly recognizable time of day. Savannah sparrows (*Passerculus sandwichensis*) can use a variety of celestial and magnetic compass cues, and use sky polarization patterns at dusk and dawn, averaging them together to calibrate their varied compass functions.

Maps

A map is a representation of physical space that allows you to determine where you are, where you want to go, and how to get there. The compass allows you to orient the map to the physical directions of the world. A process like olfactory imprinting will tell you something about your destination and allow you to recognize it once you get there. The most reliable mapping information on a global scale is the earth's magnetic field. Both long- and short-distance migrators, as well as homing animals like pigeons, use magnetic information in this way.

In addition to giving directional information through the north–south orientation of its field lines, the earth's magnetic field varies predictably over the planet in its "inclination" or "dip" angle (Fig. 5.12). The earth acts as a giant magnetic dipole (north on one end, south on the other) with magnetic field lines curving between the two poles. At each pole, the field lines are parallel to the axis of gravity; that is, they go directly down into the pole. At the equator, the field lines are parallel with the surface and perpendicular to the gravity vector. On either side of the equator, the angle between the field vector and the gravity vector, or the magnetic field's inclination, gradually changes between the two extremes. The strength of the magnetic field at the earth's surface also varies on a broad geographic scale because of variations in the internal movement of the iron core causing the field, as well as more locally because of anomalies such as iron

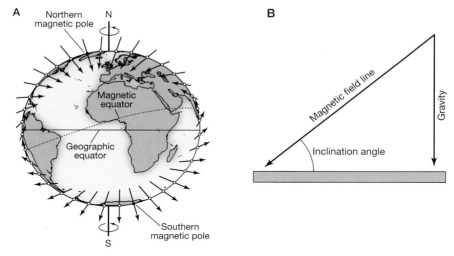

Figure 5.12. (A) The earth's magnetic field with arrows indicating orientation of magnetic field lines. (B) The angle between the field line and gravity defines the inclination or dip angle.

deposits. Magnetic-sensing animals are, in fact, very sensitive to both inclination and to field strength and use these cues as a map.

Kenneth J. Lohmann's investigations of sea turtle migration illustrate the use of magnetic information to determine geographic positions to guide navigation. Loggerhead sea turtles (*Caretta caretta*) hatch from nests on Atlantic beaches in the southern United States. They immediately enter the ocean and take up residence for several years in the current system circling between the Americas and Europe. Reliance on this system requires navigational adjustments to prevent drifting outside the boundaries conducive to foraging and to prevent moving in the wrong (northerly) direction at points where the current system splits into different streams. Lohmann tethered newly hatched turtles in a swim tank surrounded by a magnetic coil that could vary the characteristics of the field around the hatchling. He then presented magnetic field characteristics typifying different locations within their normal transatlantic routes. In each case, the swimming turtle oriented and swam in a direction appropriate to the location (Fig. 5.13).

The sea turtle work also shows the astonishing fact that both the mapping function and compass use are innate in this species. The hatchlings used in the study above and in many others are completely naive and have never encountered the particular locations and their environmental cues. How this representation is coded is a complete mystery. In other species, experience appears to be needed to learn the map. Some of the earliest and most thorough studies of magnetic navigation have employed pigeons, which will return to their home nest from whatever location to which they are transported. Homing pigeons use a variety of cues to guide them,

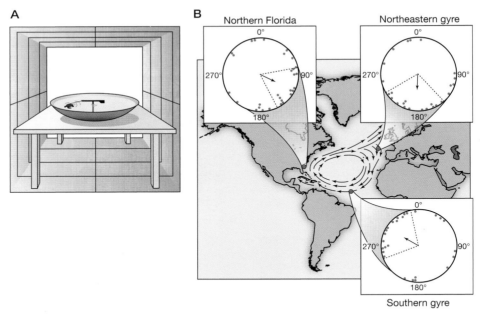

Figure 5.13. (A) Experimental apparatus used by Lohmann to test magnetic orientation in hatchling sea turtles. A turtle is suspended by a harness in a seawater tank surrounded by a magnetic coil. The magnetic field can be adjusted to mimic various geographic locations, and hatchling's swim orientation monitored. (B) Turtles placed in the apparatus with magnetic fields mimicking those of three locations in the normal migratory pattern result in swimming orientations appropriate for keeping them in the current. Red dots in the circles indicate an individual's direction; the arrow indicates the group mean direction (dashed lines = 95% confidence interval).

including olfaction, visual landmarks, and magnetic information. In one early study by Roswitha Wiltschko and Wolfgang Wiltschko, three groups of pigeons were transported from their roost to a release point several kilometers away. One group served as a control, one was placed in a magnetic coil that distorted magnetic cues only at the release point, and a third group was transported to the release site within a coil that masked or distorted magnetic cues en route. Whereas the first two groups had similar, and appropriate, headings back to the roost, the group that had been deprived of magnetic cues during their outbound trip left the release point in randomly scattered directions. Apparently, the pigeons were able to learn rapidly enough geographic information solely from magnetic variation during their outbound trip to compile a map that allowed them to know where they were in relation to their desired destination. Interestingly, older birds with more experience in homing flights were not disturbed by the magnetic interference during transport, suggesting either that their overall experience had prepared them with an understanding of geography, or that they can use different strategies when challenged in this way. It is not clear for most species whether innate or learned, or both, maps are used. However, most seasonal migrators travel routes many times over their life span, and many migrators move in groups composed of experienced and inexperienced animals. It may be that strategies change from following experienced leaders to acquiring through that experience the necessary mapping knowledge for independent travel.

The Neural Basis of Navigation

Although we speak of compasses and maps, we know nothing about how that information is represented in the brain. For magnetic sensing, it is not even clear how the nervous system represents and uses that information beyond the small amount of work on peripheral receptors. In truth, although we think of maps as the standard bicoordinate system of human paper representations of geographic space, there is no understanding of how such information is really represented in the brain. One body of work that has provided some information is that of Verner Bingman, who has investigated the role of the hippocampus in bird navigation. In Chapter 4, we noted that this brain area is involved in spatial memory and has been implicated in foraging and caching. Navigation is one type of spatial cognition involving the extraction of a sense of place and relative positions from multiple cues. Bingman found that hippocampal lesions in homing pigeons impaired their ability to use landmarks to navigate from a familiar release point to their roost. However, neither pigeons nor Savannah sparrows were impaired in their ability to orient using the sun or magnetic information, respectively. Thus, it appears that the hippocampus is not responsible for a compass function. However, as is consistent with many laboratory studies of the hippocampus's role in spatial memory, rather it participates in the cognitive processes needed for assembling and using spatial information.

CONCLUSIONS

The simple task of moving from one area to another can require particular genes and cognitive abilities for analyses of environmental stimuli as well as physiological changes that prepare the animal for movement. It can also result in a host of influences on the individual's Darwinian fitness, the population's genetic structure, and even the origin of species. We still know little of how most animals can accomplish this feat. Animal migration and orientation continue to be two of the real marvels of our natural world.

BIBLIOGRAPHY

Åkesson S, Hedenström A. 2007. How migrants get there: Migratory performance and orientation. *Bioscience* **57:** 123–133.

Audubon JJ. 1831. *Ornithological biography, or an account of the habits of the birds of the United States of America.* Adam Black, Edinburgh.

Bazazi S, Buhl J, Hale J, Anstey M, Sword G, Simpson S, Couzin I. 2008. Collective motion and cannibalism in locust migratory bands. *Curr Biol* **18:** 735–739.

Berthold P, Helbig A, Mohr G, Querner U. 1992. Rapid microevolution of migratory behaviour in a wild bird species. *Nature* **360:** 668–670.

Bingman VP, Able KP, Siegel JJ. 1999. Hippocampal lesions do not impair the geomagnetic orientation of migratory Savannah sparrows. *J Comp Physiol A* **185:** 577–581.

Bingman VP, Hough GE III, Kahn MC, Siegel JJ. 2003. The homing pigeon hippocampus and space: In search of adaptive specialization. *Brain Behav Evol* **62:** 117–127.

Buhl J, Sumpter D, Couzin I, Hale J, Despland E, Miller E, Simpson S. 2006. From disorder to order in marching locusts. *Science* **312:** 1402–1406.

Chapman J, Nesbit R, Burgin L, Reynolds D, Smith A, Middleton D, Hill J. 2010. Flight orientation behaviors promote optimal migration trajectories in high-flying insects. *Science* **327:** 682–685.

Cox G. 1985. The evolution of avian migration systems between temperate and tropical regions of the New World. *Am Nat* **126:** 451–474.

Dingle H, Drake V. 2007. What is migration? *Bioscience* **57:** 113–121.

Dittman AH, Quinn TP. 1996. Homing in Pacific salmon: Mechanisms and ecological basis. *J Exp Biol* **199:** 83–91.

Egevang C, Stenhouse I, Phillips R, Petersen A, Fox J, Silk J. 2010. Tracking of Arctic terns *Sterna paradisaea* reveals longest animal migration. *Proc Natl Acad Sci* **107:** 2078–2081.

Emlen ST. 1967a. Migratory orientation in the Indigo Bunting, *Passerina cyanea.* Part I. Evidence for use of celestial cues. *Auk* **84:** 309–342.

Emlen ST. 1967b. Migratory orientation in the Indigo Bunting, *Passerina cyanea.* Part II. Mechanisms of celestial orientation. *Auk* **84:** 463–489.

Emlen ST, Emlen JT. 1966. A technique for recording migratory orientation of captive birds. *Auk* **83:** 361–367.

Flamarique IN, Hawryshyn CW, Ha'rosi FI. 1998. Double-cone internal reflection as a basis for polarization detection in fish. *J Opt. Soc Am A* **15:** 349–358.

Gordo O. 2007. Why are bird migration dates shifting? A review of weather and climate effects on avian migratory phenology. *Climate Res* **35:** 37–58.

Gwinner E. 2008. Circannual clocks in avian reproduction and migration. *Ibis* **138:** 47–63.

Hasler AD, Scholtz AT. 1983. *Olfactory imprinting and homing in salmon: Investigations into the mechanism of the imprinting process.* Springer-Verlag, Berlin.

Hasler AD, Wisby WJ. 1951. Discrimination of stream, odors by fishes and its relation to parent stream behavior. *Am Nat* **85:** 223–238.

Labhart T, Meyer EP. 2002. Neural mechanisms in insect navigation: Polarization compass and odometer. *Curr Opin Neurobiol* **12:** 707–714.

Lohmann KJ, Lohmann CMF. 1996. Orientation and open sea navigation in sea turtles. *J Exp Biol* **199:** 73–81.

Lohmann KJ, Cain SD, Dodge SA, Lohmann CMF. 2001. Regional magnetic fields as navigational markers for sea turtles. *Science* **294:** 364–366.

Lohmann KJ, Lohmann CMF, Putman NF. 2007. Magnetic maps in animals: Nature's GPS. *J Exp Biol* **210:** 3697–3705.

McWilliams SR, Karasov WH. 2001. Phenotypic flexibility in digestive system structure and function in migratory birds and its ecological significance. *Comp Biochem Physiol A Mol Integr Physiol* **128:** 579–593.

McWilliams SR, Karasov WH. 2004. Migration takes guts: Digestive physiology of migratory birds and its ecological significance. In *Birds of two worlds: The ecology and evolution of migration* (ed. Marra P, Greenberg R), pp. 67–78. Smithsonian Institution Press, Washington, DC.

Muheim R, Phillips JB, Åkesson S. 2006. Polarized light cues underlie compass calibration in migratory songbirds. *Science* **313:** 837–839.

Phillips JB, Schmidt-Koenig K, Muheim R. 2006. True navigation: Sensory bases of gradient maps. In *Animal spatial cognition: Comparative, neural & computational approaches* (ed. Brown MF, Cook RG). http://www.pigeon.psy.tufts.edu/asc/phillips.

Ramenofsky M, Wingfield JC. 2007. Regulation of migration. *Bioscience* **57:** 135–143.

Ritz T, Adem S, Schulten K. 2000. A model for photoreceptor-based magnetoreception in birds. *Biophys J* **78:** 707–718.

Roff DA, Fairbarn DJ. 2007. The evolution and genetics of migration in insects. *Bioscience* **57:** 155–164.

Rolshausen G, Segelbacher G, Hobson K, Schaefer H. 2009. Contemporary evolution of reproductive isolation and phenotypic divergence in sympatry along a migratory divide. *Curr Biol* **19:** 2097–2101.

Semm P, Beason RC. 1990. Responses to small magnetic variations by the trigeminal system of the bobolink. *Brain Res Bull* **25:** 735–740.

Stein VW. 1977. Die Beziehung zwischen Biotop-Alter und Auftreten der Kurz-Flügeligkeit bei Populationen dimorpher Rüsselkäfer-Arten (Col., Curculionidae). *Z Angew Entomol* **83:** 37–39.

Walker MM, Dennis TE, Kirschvink JL. 2002. The magnetic sense and its use in long-distance navigation by animals. *Curr Opin Neurobiol* **12:** 735–744.

Wiltschko R, Wiltschko W. 2003. Avian navigation: From historical to modern concepts. *Anim Behav* **65:** 257–272.

Zupanc GKH. 2008. *Behavioral neurobiology.* Oxford University Press, Oxford.

CHAPTER 6

Foundations of Social Behavior

SOCIAL BEHAVIOR TAKES MANY FORMS, and the variation among animal species within those forms seems endless. Nevertheless, social behavior at its most fundamental represents an interaction between conspecific individuals driven by a curious balance of conflict and cooperation. Whether it is finding a mate, dealing with the outcome of mating (offspring), or competing for the resources necessary for reproduction, the ultimate function of social behavior is successful transfer of genes to the next generation.

Before considering specific types of social behavior, this chapter covers three topics that cut across all of them. The first is sex differences. Males and females have different physiological requirements for reproduction, and these dictate mechanistic differences, different motivations, and different selection pressures associated with social interactions. The second is seasonality. Social interactions vary throughout the year, often in ways tied to the availability of resources needed for reproduction. The third is communication. Because social behavior necessarily involves interaction between individuals, the exchange of signals between participants is necessary to coordinate the participants' behavior.

SEX DIFFERENCES

Why do males and females behave differently? This question can be answered on several levels. When posed in the context of ultimate causes, the question is an evolutionary one, and its answers lie in the differential investment by the sexes in reproduction as well as in the differential certainty of maternity versus paternity. When posed at the proximate level, the question has both ontogenetic and mechanistic considerations.

Evolution of Sex

Sex has always been a conundrum for evolutionary biologists, its reasons most clearly outlined by the eminent evolutionary biologist John Maynard Smith. Selection should favor traits that increase the passage of genes into the next generation. What if there were copies of a gene that enabled either sexual or asexual reproduction? Which variant would spread more in future generations and thus be favored by selection? Consider a sexual female who produces two offspring, one male and one female. Her daughter also produces one son and one daughter, as do all of their future descendants. All things being equal, after any number of generations, there is one male and one female and thus two copies of the allele for sexual reproduction, one in each offspring. Contrast that with an asexual female who also produces two offspring, but they are both daughters. These daughters also have two offspring, and again, both are daughters. The population of the asexuals increases from 1 to 2, 4, 8, 16, 32 to 64 in just six generations. Thus, there are 64 copies of the allele for asexual reproduction compared to just two copies of the allele for sexual reproduction after the same number of generations. All else being equal, such as males making no important contribution to offspring survivorship, the copies of genes for asexual reproduction increase at a much more rapid rate than those for sexual reproduction. This is because in asexuals all of the offspring give birth, whereas in sexual species only females give birth. This is known as the cost of males.

Given the cost of males, we wonder why there is sex. There are numerous hypotheses, but most revolve around the advantage of genetic variation associated with sexual reproduction. Sexual reproduction reshuffles the genes of the two parents into their offspring. One result is that beneficial mutations from the mother and father can be brought together, resulting in more fit offspring. Also, deleterious mutations of each parent can be brought together and result in a very unfit individual that will die but will take with it those deleterious mutations; thus, sex can edit out deleterious mutations from a lineage. Neither process can occur in asexual individuals because the offspring are clones of themselves and mutations accumulate in a ratchet-like manner, ultimately driving the asexual lineage to extinction. Hermann Joseph Muller proposed this idea as a reason for the evolution of sexual reproduction.

Sexual reproduction occurs between individuals with gametes of quite different sizes. The individual with many small gametes is the male, and the one with fewer large gametes is the female. In Chapter 7, we will discuss how these differences in gamete size result in the evolution of secondary sexual characteristics through the process of sexual selection. In this chapter, however, we are concerned with the role of primary sexual characteristics, those directly associated with the process of reproduction, and how they come about during development and maturation.

Organization and Activation of Sex Differences

What it means to be a male or a female at first seems straightforward. Females produce one type of gamete, eggs, from which an offspring will develop. Males produce another type of gamete, sperm, whose main function is to add DNA to the egg and stimulate its mitotic progression to a fully formed offspring. This basic difference between the sexes has consequences. To produce and deliver the different gametes, the two sexes need different reproductive organs, or genitalia. And to mediate the partnership that brings the eggs and sperm together, males and females supplement this basic equipment with sexually different signals and behaviors. Even beyond the immediate reproductive functions, males and females differ along almost every dimension. How do they get that way? This turns out to be a complicated question, even within a species, and even more so when considering the extraordinary range of sex-determining mechanisms and the meaning of sex across species. Here we provide a broad overview of the major components determining sexual characteristics with the understanding that there are significant departures from this model across vertebrates, and even more so when invertebrates are considered.

Sex Determination

Sex can be either genetically or environmentally determined. Genetic sex determination is the standard for mammals. As any introductory biology text reports, mammals have distinct sex chromosomes, X and Y; an individual with two X chromosomes (the homogametic sex) develops into a female, and an individual with an X and a Y (the heterogametic sex) develops into a male. Many other vertebrate and invertebrate groups employ sex chromosomes to carry the genetic markers of sex, but not all follow the mammalian pattern. In birds, for example, males are the homogametic sex. In several invertebrates, the ratio of sex chromosomes to autosomes determines sex. Amphibians are extremely diverse, in that some species within the same family can have males as the heterogametic sex, whereas in others, females are heterogametic. Other species have sex-determining genes distributed across regular chromosomes rather than on distinct sex chromosomes. In fact, much of the genetic material on X and Y chromosomes is unrelated to sex determination, and, conversely, very few genes have anything to do with determining sex. In cases in which sex-determining genes have been identified, determination derives primarily, often exclusively, from the action of a single gene. In mammals this gene, the *SRY* gene, is carried on the Y chromosome and is the key genetic difference between males and females. The *SRY* gene is responsible for triggering a hormonally mediated developmental cascade that results in a male phenotype; its absence results in development into a female phenotype.

A particularly important variation on genetic sex determination is the "haplodiploid" system found in hymenoptera (bees, termites, ants). There sex is determined by the number of chromosomes. If the individual is diploid (because of egg fertilization by sperm), the individual is female. If the individual arises from an unfertilized egg, and is therefore haploid, it becomes a male. The peculiarities that the haplodiploid system causes in genetic relationships among relatives have been argued as a key feature in the development of eusociality. This will be discussed more in Chapter 8.

An alternative to genetic sex determination is environmental sex determination, in which environmental cues early in development channel the individual into a female or male phenotype. By far the most common and well-documented influence is temperature. Temperature-dependent sex determination is found in many species of reptiles and is also seen in some fishes and invertebrates. In these cases, sex is not determined at conception, but at some point after fertilization, during a critical phase of embryonic development. The temperature difference resulting in one sex versus the other is small, often a matter of a few degrees Celsius. In some cases, lower temperatures give rise to one sex, and higher temperatures to the other. In other cases, intermediate temperatures result in one sex, whereas extremes on either end result in the other. It may seem as though temperature-dependent sex determination is unworkable in the real world, as it should lead to all males or all females depending on ambient temperature. In fact, temperature-dependent sex determination is generally found where eggs incubate in a nest where temperature extremes are buffered and vary only within a few degrees among eggs in the nest or across nests in a defined area. The small temperature variations plus the great sensitivity of the sex-determining system lead, in normal circumstances, to sex ratios of ~1:1, much as does genetic sex determination owing to the random assortment of sex chromosomes.

How temperature is translated into sexual differentiation is still not completely understood. What is known, however, seems to parallel genetic sex determination. Temperature regulates the expression of a gene (or a very small number of genes) that controls the development of steroid-producing glands such as the gonads, triggering a hormonally mediated developmental process determining phenotypic sex.

The discussion above assumes that individuals come in one of two sexes, males or females. Although that is often true, it is not the only sexual system existing in animals, and it is worthwhile to reflect on the variation in sex assignment that sexual reproduction supports. The invertebrate *Caenorhabditis elegans* is characterized by genetic sex determination based on the ratio of sex chromosomes to autosomes, but rather than determining males and females, the two sexes are males (having one X chromosome) and hermaphrodites (having two Xs). The sea slug *Aplysia* and similar gastropod mollusks depend on gene expression for normal sexual development, but all individuals are hermaphrodites, capable of donating either eggs or sperm to another individual. Even among vertebrates, different variations on sex are present. Many fish change sex; some change unidirectionally, and some are able to switch back and forth, triggered by behavioral interactions or social context. In each state, they have both the reproductive physiology and behavioral repertoire of their phenotypic sex. Whiptail lizards (*Cnemidophoris uniparens*) are an all-female species that reproduce as clones of their own unfertilized eggs. Yet each individual expresses male or female courtship and mating behavior (short of exchanging sperm), and this pseudosexual behavioral exchange seems to enhance successful, though asexual, reproduction.

The Organizational Role of Hormones

Reviewing sex determination makes the important point that the vast anatomical, physiological, and behavioral differences between males and females are not the result of a similarly vast collection of fixed genetic differences. Rather, a small genetic difference, or an environmental influence on a gene's expression, results in hormonally mediated differences in the expression of a much larger complement of genes common to both sexes. Those hormones are the sex steroids produced by the gonads—testosterone, produced by the testes in males, and estrogen, produced by the ovaries in females. Gonadal steroids function during development to guide the formation of male phenotypic traits (called "organization") as well as during adulthood to modulate the expression of male signal features and behavior (called "activation," covered in the next section). Testosterone (and other androgens) and estrogens are often dichotomized as "male" versus "female" hormones, but this is an oversimplification. It is true that in adults, testosterone secreted by the gonads regulates male-typical behavior and phenotype, whereas estrogen regulates adult female-typical behavior and appearance. However, these two classes of hormones are related by their synthetic pathways (testosterone is the immediate precursor for the synthesis of estrogen), and both sexes have measurable, and, at times, equal, circulating levels of each. Furthermore, gonadally secreted testosterone is often converted to estrogen after being absorbed by neurons to achieve its masculinizing effects on the brain. This is especially true during development, when testosterone, converted to either estrogen or another androgenic compound at its targets, has profound effects on creating a phenotypic male; but estrogen plays far less of a developmental role in creating a phenotypic female. A brief, and simplified, review of mammalian sexual development illustrates this.

As noted above, the main genetic determinant of mammalian sex is the *SRY* gene, which begins the cascade of events leading to sexual development (Fig. 6.1). This gene, found on the Y chromosome and therefore only present in males, guides production of the peptide TDF (for testes-determining factor). TDF stimulates growth and development of the testes from the embryonic Wolffian duct system, which eventually develops into the male internal reproductive system. Embryonic testes cells secrete two hormones. One is Müllerian degeneration hormone, which inhibits the growth of the Müllerian duct system, the precursor of the female reproductive tract. The other is testosterone, which is responsible for masculinizing the individual into a male phenotype. Different metabolic products of testosterone are responsible for masculinization in different body tissues. In much of the body, including precursors of the external genitals, muscles, and skeletal tissue, testosterone is metabolized by cells in those tissues to another androgen, dihydrotestosterone (DHT). DHT binds to intracellular receptors, acts as a transcription factor, and induces the developmental trajectories in those tissues, ultimately leading to the external body form and internal physiological features of a male. In the brain, however, another metabolic pathway is taken.

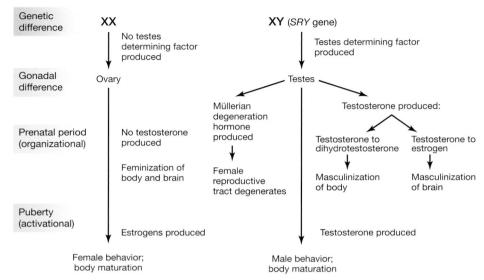

Figure 6.1. Flow diagram of sexual differentiation in mammals, which have genetic sex determination. Presence or absence of the *SRY* gene triggers the development of testes or ovaries. This, in turn, leads to a cascade of hormonal changes that lead to a male or female phenotype: Müllerian degeneration hormone (MDH) is secreted in males to block development of a female reproductive tract, whereas testosterone is (in males) or is not (in females) secreted. Testosterone, working through metabolites, causes development of male traits throughout the organism; its absence in females, coupled with the absence of MDH, leads to the default mammalian condition, a female phenotype. A second wave of endocrine release marks the beginning of sexual maturation after a juvenile period, when estrogen feminizes and testosterone masculinizes the individual's body, brain, and behavior.

Neurons convert testosterone to estrogen. Estrogen binds to intracellular estrogen receptors, which control the gene expression patterns leading to the development of a male brain. Because the brain controls behavior, this predisposes the brain to generate male-typical behavior.

Females lack Y chromosomes, and hence the *SRY* gene. As such, testes formation is not stimulated, and the Wolffian duct system degenerates. The Müllerian duct system, not subject to the testes-derived degeneration hormone, develops normally into a female reproductive system. Furthermore, without testosterone, the body and brain develop into a normal female phenotype. Although the developing ovaries secrete some estrogen, the amount is negligible compared to the high circulating levels derived from the mother. It may be the case that estrogen contributes to the feminization of the individual's phenotype to some extent, but decades of experiments show that mammals will develop into phenotypically normal females, regardless of genetic sex, as long as testosterone does not exert its influence. Females are in essence the default sex. Only the active intervention of testosterone will create a male.

Furthermore, males are a mosaic of different testosterone and genetic effects. One can see, for example, how a genetic male could develop a female phenotype because of insufficient testosterone production or sensitivity in target tissues, or an external male body developing with a female brain, or vice versa, if one of the different enzymes responsible for the two developmentally active metabolites of testosterone is defective.

Because male phenotypic development depends on levels of androgen impacting the embryo, one can also imagine individual variation in the level of "maleness" in both males and females. Classic studies in developmental psychobiology show this to be true while also highlighting how subtle developmental effects can lead to measurable differences in adult social behavior. Female rats gestate large litters, and embryonically the developing pups are lined up in the uterine horns (Fig. 6.2). Each pup's genetic sex is randomly determined by whether or not it received an X or a Y from its father. Because of this, individual females may be surrounded by two females, a male and a female, or two males. Each embryo lies in a micro-uterine environment exposed to its own hormones and the hormones of its immediate neighbors leaking into the amniotic fluid, including testosterone secreted by developing male embryos. Surprisingly, these minor variations lead to individual differences in the phenotypes of the females. Females surrounded by males in utero have slightly masculinized physical and behavioral characteristics, whereas females surrounded only by females lie toward the female end of these traits. More surprisingly, adult behavior expressed long after the uterine exposure is similarly affected. Male-surrounded females are, for example, more aggressive than female-surrounded females.

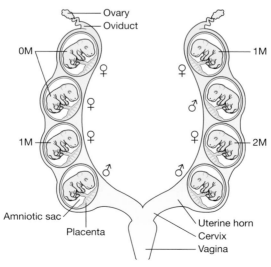

Figure 6.2. Embryonic rat pups are randomly next to zero, one, or two males. The testosterone leaking from males in utero masculinizes individuals next to them in a dose-dependent manner.

Individual differences in animals with temperature-dependent sex determin-ation can also be seen, and likely for the same reason. In leopard geckos, males are produced at higher temperatures and females at lower ones. Higher-temperature males show higher aggression than males that arise at a temperature just higher than the threshold for generating females, and have higher testosterone levels as adults. Presumably, the mechanism is similar to that in the genetically determined rats, a result of subtle downstream differences in steroid hormone levels during develop-ment, although this is still unknown. In both cases, the results show how persistent individual differences in adult behavior, differences that can significantly affect resource holding ability or reproductive success, can exist because of early prenatal hormonal effects unrelated to an individual's genetic makeup, and, as in both cases here, via totally random circumstances.

The Activational Role of Hormones

The developmental effect of hormones can organize brain centers and circuits to express male behavior preferentially. Some of this can be seen even in early juveniles in the form of sex differences in play and exploratory behavior. The major social differences between males and females, however, come with the onset of adulthood, an event in humans called puberty, wherein sexual maturity, the development of external signs of sexual identity, and adult forms of social behavior all occur at about the same time. Something must activate the latent brain circuitry and latent capacity for additional body growth organized by gonadal hormones before birth.

The first experimental study in endocrinology, performed by Arnold Adolph Bert-hold in the mid-19th century, investigated this question. Berthold removed the testes in three groups of immature male chickens. One group was not further manipulated; in the second, the birds were reimplanted with their own removed testes; in the third, the birds were implanted with testes from donor animals. Although all were geneti-cally and gonadally males as juveniles, the castrated control birds did not develop into normal adult roosters, either behaviorally or morphologically. Birds in the other two groups did, developing the normal feathers, comb, and wattles characteristic of roosters and demonstrating normal crowing and other signs of territorial displays and mate attraction.

Berthold was the first to conclude that some secretory product from the testes was responsible for the activation of systems in developmentally primed birds, allowing those latent traits to become expressed. We now know that that product, in males, is testosterone. In females, estrogen plays a similar role, activating latent female mor-phological, physiological, and behavioral traits. Thus testosterone and estrogen are, respectively, critical for the maturation of the male and female anatomical, physiolog-ical, and behavioral characteristics that mark individuals as reproductively capable adults. Before sexual maturation, circulating sex steroid levels are negligible. The basic vertebrate pattern of sex steroid levels is that they are high during some critical developmental period (in mammals, before birth), undetectable for a period of time that can be years in large mammals, then high again in adulthood (Fig. 6.3).

Once reproductive competence is reached by the activating effects of gonadal steroid hormones and animals reach adulthood, the relationship between sex, social behavior, and hormones can become complex. David Crews characterized variation in the relationship of hormones and behavior by a dichotomy of associated versus disassociated reproductive patterns (Fig. 6.4). In many organisms, reproductive and aggressive social behavior patterns are modulated by circulating sex steroids, which Crews called an associated pattern. Many animals, however, show dissociated patterns of reproductive behavior in which the behavior is not tied directly to gonadal steroid hormones. In some cases, such as zebra finches, birds are constantly ready to mate, but despite high steroid levels do not, unless some other trigger (in this case, rain) occurs. In others, such as garter snakes in high latitudes, mating occurs when both steroid levels are very low and their gonads are in a regressed winter state. Dissociated behavioral profiles tend to occur in animals living in extreme environments where breeding opportunities are either unpredictable (as in the desert-living finches) or predictable but very short (as in the garter snakes). Associated behavioral patterns typify temperate species with regular and relatively prolonged opportunities for breeding. Despite their differences, both associated and dissociated patterns require gonadal steroids to make them reproductively competent at the transition from their juvenile to adult stage of life. After that, associated species allow gonadal

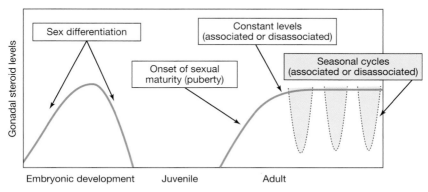

Figure 6.3. Vertebrates are characterized by two life stages of high gonadal steroids. During embryonic development, high gonadal steroids have organizing effects and influence sexual differentiation. At puberty, a second release completes sexual development and induces reproductive capability, a combination of organization and activation effects. After that, adults can have steady or cyclic patterns of hormone release, and behavior can be associated or disassociated with the levels of hormone (i.e., be activated by it). Between these periods, hormone levels are undetectably small.

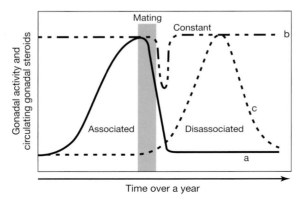

Figure 6.4. Adult patterns of circulating gonadal steroid hormones vary. Species can have an associated pattern (a), in which behavior tracks the rise and fall of hormone level, or one of two dissociated patterns: (b) one in which a constant high level of hormones is not enough to induce behavior, and another, often environmental, cue is necessary to enable the behavior; or (c) behavior occurs in the absence of circulating hormones, although hormones may rise later to induce physiological reproductive activation.

steroids to modulate their behavior, whereas dissociated species do not, although each still has high hormone levels at least some part of the year.

The role of gonadal steroid hormones in organization and activation is a paradigm that applies to vertebrates, but invertebrates are far more complex and diverse in their life history patterns. Nevertheless, the general pattern holds in many cases. Hormonal substances of some kind are important in guiding development (organization) as well as activating and often modulating behavior in later life stages.

SEASONALITY

It is a rare habitat on this planet in which resources remain constant over a year. Organisms must deal with this by managing both their energy needs and predicting optimal times to express energy-intensive functions. This is the nature of seasonality, seen as cyclic changes in behavior and physiology tracking environmental conditions and cues. Seasonality is observed in most areas of animal behavior, including foraging and migration (Chapters 4 and 5). However, reproduction, which is both energy intensive and critical from a selective standpoint, is a primary driver of seasonal variation in social behavior. From an animal's point of view, the key to optimal matching of behavior to the environment is matching an internal regulatory mechanism with an external environmental signal that indicates or predicts the presence or impending absence of needed resources. There are three ways in which animals do this: using the presence or absence of the resource itself to trigger behavioral shifts, using a reliable environmental cue that coincides with available/absent resources to trigger the shift, or evolving an endogenous circannual rhythm in internal state that generally parallels an annual environmental cycle.

Seasonal Patterns and Environmental Triggers

Linking behavioral change directly with resource change is the most direct method for seasonality. Tropical habitats, for example, often have distinct, prolonged wet and dry seasons. Most tropical frogs need standing water to breed, and the onset of rain triggers the onset of their breeding behavior, resulting in the cacophony of frog calls in the tropical wet season.

Triggering a seasonal change directly by the needed resource is reliable but provides no basis for predictive planning. Consider the problem of a large mammal with a long gestation period. It is advantageous to plan mating so that birth occurs at a time of mild temperatures and abundant food. Similarly, consider a small mammal or bird that needs to predict the lack of food well in advance, so that fat stores can be increased before that time while food remains available. These are situations commonly faced by temperate zone animals. There it is advantageous to synchronize behavioral state with environmental cues that predict the availability of resources. The most obvious cue is day length, or photoperiod (the ratio of light and dark during the 24-h daily cycle). Small animals that mate, gestate or brood, give birth or hatch, and have young that can reach at least limited self-sufficiency quickly are "long-day breeders." Lengthening days (or, more specifically, an increasing ratio of light to dark periods per day) trigger reproductive social behavior. In this way, young are produced while conditions are mild and food is abundant. Large mammals like deer or sheep are "short-day breeders." Here, a decreasing ratio of light to dark triggers reproductive social behavior. These animals thus breed in the fall, females remain pregnant during the winter, and then birth occurs in the spring, when resources become abundant and will remain so for several months, ensuring optimal conditions for the young to survive. Note that it is not universally true that a particular day length dictates reproduction. Rather, day length is simply a predictor of where an animal is in the annual cycle of resource abundance and shortage. Each species must, in an evolutionary sense, then compare that point to when the optimal time for offspring survivorship will occur and so time mating behavior appropriately.

Other seasonal behaviors also make use of the predictive power of the 24-h circadian cycle. As discussed in Chapter 5, seasonal migration is a way of following resources. Shortening daylight periods predict that food in your area will gradually decrease, triggering the behavioral and physiological shifts leading to migration, whereas lengthening daylight periods signal that it will soon be safe to return. Figure 6.5 shows the cyclic changes in environment, behavior, and physiology in a

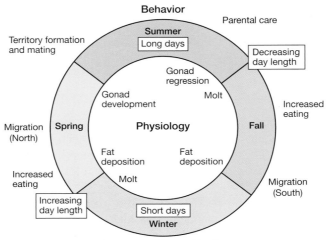

Figure 6.5. Hypothetical seasonal cycle in physiology and behavior in a temperate zone bird that migrates within the northern hemisphere. High gonadal hormones during the long-day period of spring and summer induce reproductive behavior. As days decrease, gonads regress, gonadal steroid levels drop, and other physiological events trigger molting, increased feeding, and fat deposition to prepare for fall migration. Social behavior in the nonreproductive, short-day winter period is completely different from the aggressive territorial and courtship behavior, with feeding flocks apparent. The change in day length at the end of winter triggers physiological events similar to those in the fall, leading to the spring migration bringing the bird back to the summer breeding grounds.

hypothetical northern hemisphere bird. Short days are also a cue for hibernating mammals or other species that show winter torpor to shift into an energy-saving mode to survive the coming scarcity of food.

Using photoperiod as a trigger for seasonal changes in behavior is obviously not useful when there is little or no annual variation in day length. Tropical habits are one example where photoperiodism is less important than direct responses to changes in rain or food availability. Another situation where photoperiod is of little use in triggering the seasonal emergence from winter torpor is in burrowing or other sequestered species. Many such species have evolved an endogenous circannual rhythm; that is, seasonal changes are triggered by some internal clock that cycles on a yearly basis. In laboratory experiments, ground squirrels (*Spermophilus lateralis*) display periodic breeding activity about every 350 d (individuals vary from 330 to 380 d), and also show approximately yearly cycles of food intake, body mass, and torpor when kept in constant light and temperature conditions. Exposure to short or long days can, however, entrain this endogenous rhythm, allowing the rhythm to match natural environmental conditions and synchronize individuals with different endogenous rhythms to the same "real" yearly cycle of resource availability. Between animals locked to environmental cues and animals with truly endogenous rhythms, there are a large number that are sensitive to day length but become insensitive at critical periods or after some period of time in constant conditions. This allows the spontaneous emergence of reproductive physiology and behavior after a period of hibernation.

Hormonal Mediation of Seasonality

Seasonal changes in social behavior are mediated by seasonal changes in gonadal steroids in species with an associated reproductive pattern (see above). In male deer, for example, the short days of fall trigger elevations of testosterone. The rising androgens foster the growth of external signs of reproductive state (antlers) and initiate associated behaviors such as bellowing and courtship. This same pattern is seen in spring-breeding mammals and birds, but there long days trigger the androgen rise, and in the case of species with endogenous patterns, a spontaneous elevation in gonadal steroids occurs to enable the physiology and behavior necessary for reproduction. Increases in estrogen in females, triggered by the same environmental cues elevating testosterone in males, result in an increase in female receptivity and external morphological, chemical, or behavioral signs of fertility. Gonadal steroids thus act as the common coordinator for reproduction from the production of gametes to the expression of behavior that leads to their exchange. Animals with dissociated reproductive patterns do not directly link behavior with hormone levels. Nevertheless, in many such species, a circannual change in gonadal steroids still occurs to regulate the physiological capacity for reproduction, even if this occurs temporally distinct from the behaviors that bring the males and females together (Fig. 6.4).

COMMUNICATION

Communication, at its most basic, is the process by which one individual, the sender, produces signals that have the potential to influence the behavioral response of another individual, the receiver. Signals are aspects of the sender's phenotype that evolved for the function of communication. These signals can interact with the receiver in different sensory modalities, such as through the auditory, visual, or olfactory domains.

One can view communication as a cooperative venture in which the sender and receiver exchange information to the mutual benefit of each interactant. This is more likely to be true when the sender and the receiver have common interests. Alternatively, in the more common scenario in which there is a conflict of interests, communication is an efficient means to manipulate the behavior of another individual. Instead of using physical force to compel another to provide food, to flee, or to mate, a sender merely produces a signal that can have these effects. For example, it is beneficial, in the sense of Darwinian fitness, for a mother bird to feed her nestling in response to its begging display. But the mother's interest is in the survival of all of her nestlings, whereas an individual is more interested in its own survival. Thus, an individual nestling will continue to beg to receive a larger portion of the food the mother has to deliver even if this is to the detriment of its nest mates. In this example, the begging display might not be a reliable or honest predictor of how nourished the nestling is, only of how much food it wants. It is this conflict of interest that results in many communication systems evolving like an arms race, with the sender evolving ways to manipulate the receiver's behavior, and the receiver evolving mechanisms to guard it from such manipulation and to respond to the signal in a way that benefits itself rather than the sender. In Chapter 7, we will discuss various ways in which a signal is kept "honest"; that is, it reliably predicts something about the quality of the sender.

From Signal to Receiver: The Intervening Environment

Once a signal is produced, it must travel through the environment to reach the receiver. When communication takes place over long distances, the signal that reaches the receiver can have little resemblance to the signal that was produced by the sender. In acoustic communication, for example, the signal decreases in both amplitude (attenuation) and quality (degradation) over distance. Selection should favor the sender to increase the active space of its signal, the area over which the signal is perceived by the receiver. There are a variety of adaptations that allow this.

Animals can evolve signal features that suffer less degradation and attenuation. Bird songs, for example, evolved in response to the acoustics of their local habitat. In a classic study, Eugene Morton showed that the dominant frequency of song in birds in Panama varies predictably among habitats. In general, lower frequencies travel better on the forest floor. Birds that reside on the forest floor produce the lowest frequencies, whereas those in the canopy, the forest's edge, and grassland produce progressively higher frequencies. Habitat also affects temporal patterning in bird song; cluttered habitats cause smearing of pulses and thus loss of temporal information. Great tits are a common bird in Europe and are found in a variety of habitats there. Birds in forests produced songs that were tonal and of lower frequency, whereas those in open habitat produced songs with more temporal patterning (Fig. 6.6). Another modern source of interference with long-distance communication is anthropogenic noise. Urban environments are characterized by a low-frequency din generated by various industrial applications. In urban areas, great tits use higher-frequency call components that allow their song to better contrast to this background noise.

Signal–Receiver Adaptations for Active Space

A signal's active space is also influenced by the sensitivity of the receiver. As we discussed in Chapter 3, receivers vary in the stimuli to which they are most sensitive. Ears and eyes, for example, are only sensitive to a subset of the acoustic and visual universe and are more sensitive to certain wavelengths within their range of sensitivity.

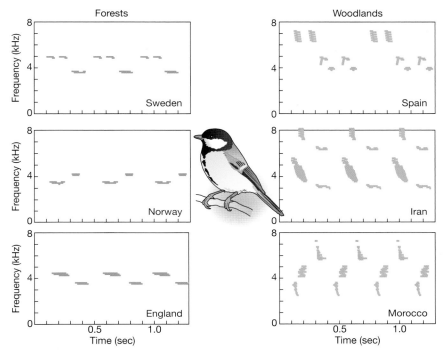

Figure 6.6. The songs of great tits vary as a function of habitat. Sonograms show that songs produced by birds residing in the forest are more tonal, have a more restricted bandwidth, and less temporal patterning than songs produced in woodlands, which are a more open habitat.

The interaction between signal and receiver, along with the effect of the environment, is seen quite clearly in the cichlid fishes of Lake Victoria. When light travels through water, there is excessive attenuation of the short wavelengths of their visible spectrum. Thus, at deeper sites, and sites at the same depth but with more particulate matter in the water, the ambient light is reshifted. Ole Seehausen and his collaborators have shown that fishes that normally have red or blue coloration tend to lose this color in deep and polluted sites (Fig. 6.7A,B). There are also changes in the receiver as a function of the ambient light environment. There is substantial sequence variation in the long-wavelength (red)–sensitive cone in these fishes. Emerging from this variation are two major photopigment alleles that differ by ~15 nm in peak sensitivity. The blue-shifted photopigment (P) is more common in populations dominated by the blue male phenotype, which is always in shallower water. There is an increase to the red-shifted (H) phenotype at greater depths, where the red phenotype is more common (Fig. 6.7C). In populations in which all males are in shallow water and are blue, P is the common allele. Other populations that range over a greater depth have males that are both blue and red. These populations are characterized by the

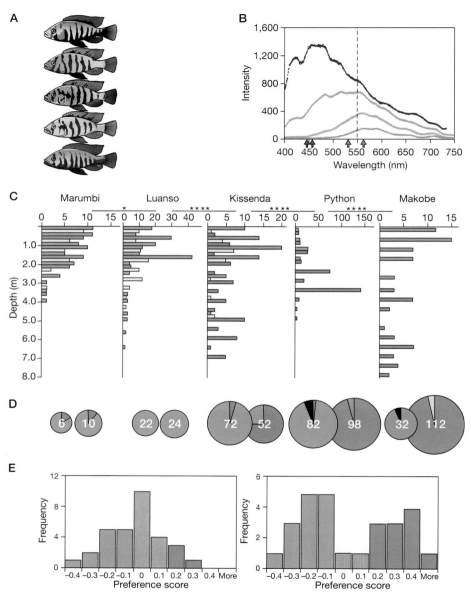

Figure 6.7. The relationship between male color, ambient light, opsin gene sequence variation, and female mating preferences in Great Lake cichlids. (*A*) Male cichlids vary substantially in nuptial coloration. The figure illustrates five phenotype classes from (*top*) 0 (blue, typical *Pundamilia pundamilia*) to (*bottom*) 4 (red, typical *Pundamilia nyererei*). (*B*) The amount and quality of ambient light vary substantially across water depth. The figure illustrates an example of a moderately steep light gradient (Python Island): surface light spectrum (blue) and three subsurface light spectra measured at 0.5 m (green), 1.5 m (light orange), and 2.5 m (dark orange) water depth. (*See facing page for legend.*)

presence of both P and H alleles (Fig. 6.7C). The mating preference also maps onto these patterns of variation in color and color sensitivity. In Luanso Island, a shallow population, most females have preferences for the blue phenotype (Fig. 6.7F, left), whereas at Python Island there is a polymorphism in the mating preferences, just as there are polymorphisms in the male color morph and the photopigment alleles.

Signal – Receiver Adaptations for Active Time

Many animal signals are ephemeral. The pushup display of a lizard and the roar of a lion are there and gone. But sometimes they are remembered. Although studies in animal communication have addressed the importance of active *space*, there has been less consideration of the parallel aspect of active *time*. Just as signals are likely to be more effective if they encompass greater space, they should also prove more effective if they encompass greater time.

Karin Akre and her colleagues investigated how complexity of mating calls of the túngara frog influence receiver memory. The frogs have a simple call, the whine, which can be followed by 0–7 chucks. Females prefer calls with chucks to calls without chucks. These frogs, like many other species of frogs and insects, engage in chorusing bouts in which there is a substantial silent period between calling bouts. In túngara frogs, this silent period is on average ~25 sec. Female frogs are tested in phonotaxis experiments in which they approach to within a few centimeters of a

---

Figure 6.7. *(Continued)* The line through 550 nm indicates the divide used to calculate the ratio of shorter- to longer-wavelength light. The arrows indicate peak absorbance or sensitivity of two opsin pigments: the two main allele groups at the long-wavelength-sensitive opsin locus (544 nm and 559 nm) and two arrows indicating the known range of peak absorbance at the short-wavelength-sensitive locus. (C) The distribution of male nuptial color phenotypes varies with depth. (Blue bars) Blue-colored males; (pale yellow bars) intermediate; and (orange or red bars) red males (orange if dominated by class 3; red if dominated by class 4). Significance levels of differences between islands in the divergence between red and blue: (*) $p < 0.05$, (****) $p < 0.001$. (D) The sequences of the opsin genes vary among sites. Here we see the frequencies of functional allele groups at the long-wavelength-sensitive opsin gene by island and male color. (Blue) Alleles of the P group; (red) alleles of the H group; (yellow) M3 alleles; (black) other alleles. On the *left* for each island are shown the opsin genotypic classes for blue-colored males, and the *right* pie chart shows the opsin classes for red-colored males. Numbers report sample sizes of completely sequenced haplotypes. For Marumbi Island and Luanso Island, only the haplotypes of those individuals are included that could be assigned to blueish and reddish phenotypes (altogether 24 and 54 haplotypes were sequenced from Marumbi and Luanso, respectively). Fish from Marumbi were divided into classes 0–1 and classes 2–3. Fish from Luanso were divided into classes 0–1 and 2–4. At all other islands, only fish of phenotype classes 0 and 4 were included. (E) Females from two populations were tested in five choice tests each to determine their strength of preference for blue versus red males. The negative scores and blue bars represent a preference for blue, the intermediate scores and gray bars no preference, and the strongly positive scores and red bars a preference for red males. Preferences for females from Luanso Island (*left*) and Python Island (*right*) are shown.

speaker broadcasting a mating call; when females are given a choice between two calls, this method determines the female's call preference. In these experiments, a female restrained in the center of the arena was broadcast a whine from one speaker and either a whine with one chuck or a whine with three chucks from another speaker. There was a silent period that was then followed by whines from each speaker. If a female remembered the speaker that broadcast the call with the chuck, she should approach that speaker. If not, she should be as likely to approach either speaker. In these experiments, females exhibited memory for the whine with three chucks over a 45-sec delay period, which is more than the typical silent period between chorus bouts. There was no memory for the call with one chuck. Thus, one way to make the call more memorable in this species to increase its active time is to produce a call with three chucks.

Animal Communication and a Human Metaphor: Language or Music

Both animals and humans behave, and it is tempting to invoke anthropomorphism to explain how and why animals behave. Consider the orca of Sea World fame. The pattern of white coloration below the lower mandible and the black coloration above it resembles a human smile, which is endearing to so many tourists at marine parks. But it belies the orca's other name, killer whale, which more accurately reflects its position in the natural world as a top predator that feeds on seals, walruses, and whales. Anthropomorphism can be deceiving.

Most evolutionary biologists would argue that we are a product of descent with modification, as are all the organisms on this planet. But it has always been contentious to apply Darwinian logic to human behavior. Many in the social sciences, and even some in the natural sciences, feel that culture rather than biology explains more regarding who we are and what we do. Of course, the dichotomy is false; culture is as much of our biology as are our genes. For example, is song learning in birds and the dialects that result from it not a valid concern of the biological sciences?

In most areas of behavioral science, animal behavior has been used to expand our understanding of the mechanisms and evolution of human behavior in the spirit of any species comparison. The tables are reversed in an interesting way when it comes to animal communication: Sometimes our understanding of the animal behavior we observe is constrained by how well it fits our understanding of human language. Humans use language to communicate. Merely the title of the endeavor, "animal communication," suggests that animals do the same, and thus the metaphor of language has been applied extensively, either explicitly or implicitly, to animal communication. The notion sometimes seems to suggest that to understand animal communication, researchers need to become Dr. Dolittles and just translate what all these animal signals mean.

To apply a language-based analysis to animal communication requires several assumptions. The central emphasis is on what signals mean, and must assume some parity in representation of signal meaning: The sender and the receiver both know the meaning of a signal. The design of the signal is not critical, as its structure is arbitrary relative to the signal's meaning. This leads to an even more demanding assumption, that communicators have a theory of mind; they must assign mental states, such as belief systems, knowledge, desires, and intents, to other individuals. There is no question that this happens with us, but there is no evidence that it happens with most of the other animals that communicate.

The strongest link between animal communication and human language has been in the classic studies of referential signals in vervet monkeys by Peter Marler, Robert Seyfarth, and Dorothy Cheney. Individual monkeys have different alarm calls for different types of predators (e.g., snakes, hawks, and leopards). These signals seemed similar to human language in that the assignment of the signal to the predatory situation is arbitrary. However, some of these same researchers later showed a lack of intentionality on the part of the sender and pointed out that in this aspect the vervet communication system is fundamentally different from human language. Given this caveat, their warning system is usually now referred to as "functionally referential."

It is important to understand cognitively complex communication behavior in nonhuman animals that might provide insights into the biological basis of our capacity for language, and the vervet system is paramount among them. But adopting human language as the benchmark to which all animal communication systems should be compared is apt to lead researchers astray. Consider songbirds, frogs, and crickets incessantly repeating the same courtships signals again, and again, and again, thousands of times per day or night. Should we really think of them as repeating the same sentences with the same words ad nauseam? Are the senders trying to inform receivers with some message, or, as Richard Dawkins and John Krebs famously suggested, are senders trying to manipulate the behavior of receivers and receivers trying to resist such manipulations when it is to their advantage to do so? If animal signaling is more about manipulating noncognitive responses of receivers—such as attention, motivation, and hormonal and emotional state— perhaps a better metaphor for animal communication is music, where the sensory pattern is not meant to be analyzed note by note by the receiver but, rather, is constructed by the sender to hold the receiver's attention and induce a more subjective response.

In summary, most instances of animal communication occur between individuals with conflicts of interest. Evolutionary theory predicts an arms race, in which senders evolve signals that manipulate the behavior of receivers and receivers evolve responses that promote their own interests. One example of such manipulation of receiver behavior is the example in Chapter 1 of how the swords of swordtails exploit preexisting female mating preferences. In the following chapter on mate choice, we further examine how the receiver's sensory, neural, and cognitive biases define the types of signals that senders evolve.

CONCLUSIONS

Sex differences, seasonality, and communication lie at the heart of social behavior. In fact, they are germane to virtually all behavior, including the resource-related behaviors of foraging and migration discussed in earlier chapters. Sex differences are especially important to understanding biological patterns. They inform the understanding of all four of Tinbergen's four questions—evolution, function, mechanism, and development. Seasonal patterns of social behavior are strongly linked to seasonality in a host of other behavioral, physiological, and anatomical characteristics that vary cyclically in time. Communication has a special relationship with social behavior and could be considered to define it, regardless of whether the communication and the social interactions it supports occur in the context of social foraging, migration, reproduction, or aggression.

BIBLIOGRAPHY

Akre KA, Ryan MJ. 2010. Complexity increases working memory for mating signals. *Curr Biol* **20:** 502–505.

Becker JB, Breedlove SM. 2002. Introduction to behavioral endocrinology. In *Behavioral endocrinology* (ed. Becker JB, et al.), pp. 3–38. MIT Press, Cambridge, MA.

Becker JB, Breedlove SM, Crews D, McCarthy MM, eds. 2002. *Behavioral endocrinology*. MIT Press, Cambridge, MA.

Breedlove SM, Hampson E. 2002. Sexual differentiation of the brain and behavior. In *Behavioral endocrinology* (ed. Becker JB, et al.), pp. 75–111. MIT Press, Cambridge, MA.

Cheney DL, Seyfarth RM. 2005. Constraints and preadaptations in the earliest stages of language evolution. *Linguist Rev* **22:** 135–159.

Crews D. 2002. Diversity and evolution of hormone-behavior relations in reproductive behavior. In *Behavioral endocrinology* (ed. Becker JB, et al.), pp. 223–284. MIT Press, Cambridge, MA.

Crews D, Bull JJ. 2009. Mode and tempo in environmental sex determination in vertebrates. *Semin Cell Dev Biol* **20:** 251–255.

Dawkins R, Krebs JR. 1978. Animal signals: Information or manipulation. In *Behavioural ecology: An evolutionary approach* (ed. Krebs JR, Davies NB), pp. 282–309. Blackwell Scientific, Oxford.

Hunter ML, Krebs JR. 1979. Geographical variation in the song of the great tit (*Parus major*) in relation to ecological factors. *J Anim Ecol* **48:** 759–785.

Krebs JR, Dawkins R. 1984. Animal signals: Mind-reading and manipulation. In *Behavioural ecology: An evolutionary approach* (ed. Krebs JR, Davies NB), pp. 380–402. Blackwell Scientific, Oxford.

Maynard Smith J. 1978. *The evolution of sex*. Cambridge University Press, Cambridge, UK.

Morton ES. 1975. Ecological sources of selection on avian sounds. *Am Nat* **109:** 17–34.

Muller HJ. 1932. Some genetic aspects of sex. *Am Nat* **66:** 118–138.

Nelson RJ. 1995. *An introduction to behavioral endocrinology*. Sinauer, Sunderland, MA.

Nelson RJ, Badura LL, Goldman BD. 1990. Mechanisms of seasonal cycles of behavior. *Ann Rev Psychol* **41:** 81–108.

Rendall D, Owren MJ, Ryan MJ. 2009. What do animal signals mean? *Anim Behav* **78:** 233–240.

Sakata JT, Crews D. 2003. Embryonic temperature shapes behavioural change following social experience in male leopard geckos, *Eublepharis macularius*. *Anim Behav* **66:** 839–846.

Seehausen O, Terai Y, Magalhaes I, Carleton K, Mrosso H, Miyagi R, van der Sluijs I, Schneider M, Maan M, Tachida H. 2008. Speciation through sensory drive in cichlid fish. *Nature* **455:** 620–626.

Seyfarth RM, Cheney DL, Marler P. 1980. Monkey responses to three different alarm calls: Evidence of predator classification and semantic communication. *Science* **210:** 801–803.

vom Saal FS. 1979. Prenatal exposure to androgen influences morphology and aggressive behavior of male and female mice. *Horm Behav* **12:** 1–11.

CHAPTER 7

Species Recognition, Mate Choice, and Sexual Selection

O NE OF THE MOST IMPORTANT DECISIONS an animal makes is choosing an appropriate mate by assessing an individual's sex, species, health, vigor, and even its genetic constitution. This assessment can take place in all of the sensory modalities and can invoke complex signal analysis and decision making. Much of mate choice involves communication between a male and a female. Often the male presents courtship signals, and the female assesses those signals. The details of how these traits and preferences evolve relative to one another have been of interest to evolutionary biologists for many years.

SPECIES RECOGNITION, MATE CHOICE, AND SPECIATION

There was a revolution in the biological sciences in the mid-20th century when the theory of evolution was revitalized in the modern synthesis, which brought together data from fields including population biology, genetics, paleobiology, and behavior. The biological species concept, put forth by Ernst Mayr, emerged from the synthesis and defined a species as a group of interbreeding or potentially interbreeding organisms. Although there are other definitions of species and problems with all of them, Jerry Coyne and H. Allen Orr argue in a recent review that the biological species concept is the one that has stood the test of time.

A cornerstone of the synthesis is the processes by which species multiplied, and this is where the contribution of behavioral studies was most prominent. When reproduction is disrupted through geographical barriers between populations, gene flow can cease and the populations become more and more different from one another over time, either because of genetic drift or because the populations evolve in response to local selection pressures. Behavioral mechanisms such as mate choice can contribute to the process of speciation, and in some cases may be the final coup de grâce. For example, when populations diverge, they might become different enough to not recognize each other as potential mates and cease interbreeding if the two populations later merge. If mate recognition does not occur, then breeding ceases or is severely limited, with the consequential further reduction of gene flow and the acceleration of divergence between the populations, until at some point they evolve into different species—speciation has taken place. As can be seen from this point of view, reproduction is key to understanding species, and mate recognition is key to reproduction.

Species Recognition

Mate recognition between species is often called species recognition. The latter term should be considered a description of mate recognition when it happens between members of different species, but it should not be assumed that species recognition has evolved to ensure the genetic integrity of the species. Species recognition occurs, but it can be a consequence of mate recognition at other levels.

There is strong selection for individuals to avoid mating with heterospecifics, and when such matings occur, fertilization often does not take place. If it does take place, development often goes awry, and if hybrid individuals are born, they often have reduced vigor. Some recent studies have shown that hybridizations between species are sometimes advantageous, but this is clearly the exception and not the rule.

When heterospecific matings do occur, the cost is greater for females than for males. This is because females can usually mate only one or a few times in each breeding season, whereas males can often mate continuously. As we will discuss further below, it is usually the male that conspicuously advertises for females and the female that exercises choice among the males. As a result, female mate recognition and male courtship traits have been the focus of species recognition. All species that reproduce sexually have species-specific courtship traits—that is, traits in which the variance is much smaller within the species than among species. Such traits occur in all modalities and are exemplified by pheromones in moths, color patterns in fishes, and songs in birds. Species-specific traits are complemented by species-specific recognition—that is, recognition of only those individuals bearing those traits as appropriate mates.

Reproductive Character Displacement

The best illustrations of how selection influences the evolution of species mate recognition systems are in the context of reproductive character displacement. This occurs when differences in species-recognition signals or preferences between species that reside in different areas (allopatry) are enhanced when those species occur together (sympatry), with the result that heterospecific matings or mate recognition errors are reduced. A recent example comes from studies of two species of chorus frogs in the genus *Pseudacris*, *Pseudacris nigrita* and *Pseudacris feriarum*, by Emily Moriarty Lemmon.

Chorus frogs, like almost all frogs, produce species-specific mating calls. These calls are used by females to assess the male's appropriateness as a mate, both in comparisons of conspecific versus heterospecific, and among potential conspecific partners. The call of the chorus frogs resembles the sound of dragging a finger across the teeth of a comb, a series of rapid trills (Fig. 7.1). For many of these kinds of frogs, as well as for crickets, trill rate is a key species-recognition attribute. In allopatry, the *P. nigrita* call has fewer pulses and a slower pulse rate than does the *P. feriarum* call.

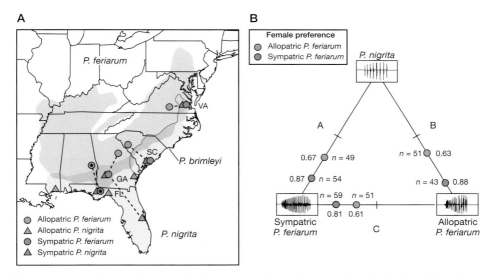

Figure 7.1. Reproductive character displacement in chorus frogs. (A) The distribution of the chorus frog *Pseudacris feriarum* and *Pseudacris nigrita* in the southeastern United States and populations sampled. Call transects are indicated by dashed lines. Abbreviations are for states (VA, Virginia; SC, South Carolina; GA, Georgia; FL, Florida). The distribution of *Pseudacris brimleyi*, another closely related chorus frog, occurs within the area defined by the solid gray line. Female phonotaxis experiments were conducted with frogs from the sites indicated by a black star. (B) A summary of female phonotaxis results from the three experiments in which female *P. feriarum* from sympatric and allopatric sites were given a choice between (A) calls of sympatric male *P. feriarum* versus *P. nigrita*, (B) calls of allopatric male *P. feriarum* versus *P. nigrita*, and (C) calls of allopatric versus sympatric *P. feriarum*. An oscillogram of a natural call from the respective populations is shown at each vertex of the triangle. The proportion of females that chose the more popular stimulus is indicated by its position along the line segment connecting the two populations from which the mating calls were tested. The proportion of females preferring the most preferred call and the samples sizes for each experiment are also indicated. Black tick marks on the line segment indicate the null expectation under no preference (0.50).

There is reproductive character displacement in sympatry. At three sympatric sites, there was displacement in the *P. feriarum* call, which had faster and/or more pulses than it did in allopatry, and at sympatric sites, there was displacement in the *P. nigrita* call, which evolved to be slower and have fewer pulses than it does in allopatry. Female phonotaxis experiments in frogs and insects present a female with different calls; the female's call preference, and by extension her mate preference, is indicated by her approach to one of the calls. Phonotaxis experiments with the chorus frogs provided evidence for evolution of mating call preferences in sympatry as well. Female *P. feriarum* frogs from allopatric and sympatric populations were given a choice between conspecific versus heterospecific calls from sympatric and allopatric populations. In all cases, females preferred the conspecific calls to the heterospecific calls (Fig. 7.1). Females of both species at the sympatric site, however, were more discriminating because they exhibited a stronger preference for the conspecific call compared to both the allopatric and sympatric *P. nigrita* calls (Fig. 7.1).

This study also shows how mating preferences between species can be inextricably linked to mating preferences within the species, a link we will return to in our discussion of sexual selection. As might be expected, female *P. feriarum* from sympatry preferred the mating calls of sympatric *P. feriarum* to allopatric *P. feriarum*. But unexpected was the preference of allopatric *P. feriarum* for the sympatric conspecific call over the conspecific call in their own population. It seems that the female preference in sympatry has driven the evolution of the male call in a direction, toward faster and/ or more pulses, already preferred by females from both allopatric and sympatric sites. Thus, sexual selection for call parameters within the species becomes intertwined with the evolution of call differences between species. Besides illustrating the action of reproductive character displacement on mating calls and preferences, this study shows how a change in mating signals can simultaneously cause outcomes that are typically classified separately as species recognition and conspecific mate choice (i.e., sexual selection). Both are consequences of the interaction of signal and receiver, though it is not easy to tell which is the evolved function.

NEURAL MECHANISMS GUIDING MATE RECOGNITION

The neural mechanisms underlying the recognition of conspecific mating signals are best understood in the auditory processing of frogs and crickets. In both cases, males produce an acoustic signal, and females base their mate choice on those signals.

Robert Capranica began the neuroethological approach to acoustic communication through his work on the bullfrog (*Rana catesbeiana*), summarized in a classic 1968 paper with Lawrence S. Frishkopf and Moise H. Goldstein, Jr. Just as Jerome Y. Lettvin asked what the frog's eye tells the frog's brain when considering the natural behavior of prey capture (see Chapter 4), Capranica asked what the frog's ear tells the frog's brain in the context of mate-calling recognition. Capranica proposed the "matched filter hypothesis," which posited that call recognition is supported by a match between the spectral, or frequency, characteristics of the signal and the filtering properties, or tuning, of the ear. The ear is thus most sensitive to frequencies in the conspecific call while filtering out (being less sensitive to) frequencies outside it, including those that may characterize the calls of heterospecifics. In this way, detecting the call is facilitated in noisy environments, and recognizing it is enhanced by the nervous system listening for a particular pattern of neural activation across the sensitive (matched) areas of the ear's receptor system that is typical of the conspecific call.

Capranica used the bullfrog call and male responses to it as an example of call recognition via a match filter mechanism (Fig. 7.2). Like all frogs and toads, bullfrogs have two auditory receptor organs in the inner ear with different physiological properties. The amphibian papilla (AP) has receptors sensitive to lower-frequency sound and has two distinct populations of receptors, a low-frequency (in bullfrogs and most other frogs with peak sensitivity $\sim$300 Hz) and a mid-frequency population (with peak sensitivity $\sim$700 Hz). Furthermore, activation of the mid-frequency population suppresses activation of the low-frequency cells. The basilar papilla (BP) is anatomically and physiological distinct and is tuned to frequencies higher than the mid-frequency receptors of the AP ($\sim$1500 Hz in bullfrogs, but varying widely across amphibian species). The bullfrog advertisement call has peak acoustic energy matching the low AP and the BP receptors, and the simultaneous activation of these two populations by the call or a reduced, synthetic signal evokes a calling response in the listener. Equally important, simultaneously activating the mid-frequency AP population causes a suppression of evoked calling: Such would be the case if the frog were being stimulated by broadband noise rather than by the discrete frequency composition of the call. Capranica also noted that the bullfrog distress call has a single frequency peak matching the mid-frequency sensitivity peak, and hearing the distress call also suppresses evoked calling. Remember that the mid-frequency AP population suppresses activity of the low-frequency AP cells; thus mid-frequency AP activation would block the nervous system from detecting the twin-frequency peaks of the bullfrog's call. The matched filter hypothesis and its relevance to both male and female responses to acoustic signals have been generally supported in numerous amphibian species. For both the spectral composition and the temporal pattern, call characteristics generally match areas of emphasized sensitivities of the ear.

In his 1968 paper, Capranica speculated that there should be call detectors somewhere in the brain that respond to the call's combination of frequencies and temporal properties that the ear processes separately. As we have seen in other systems, neural response properties become more complex at each stage of the central nervous system. The midbrain auditory center (called the torus semicircularis in frogs and the inferior colliculus in mammals) does, in fact, contain neurons that respond preferentially to various frequency combinations or have particular sensitivity to temporal properties such as pulse rate that are characteristic of the conspecific call. Just as Capranica noted for the ear, a kind of matched filter process occurs in the midbrain in

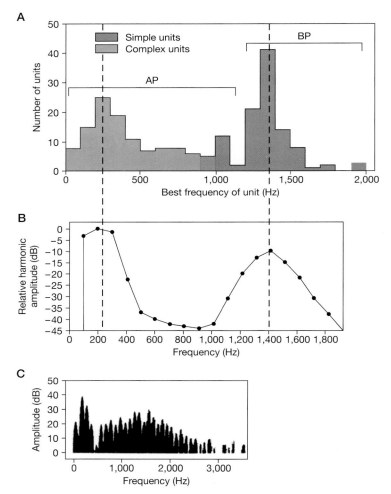

Figure 7.2. Neural mechanisms of call recognition in bullfrogs. (*A*) Distribution of auditory neurons in the VIIIth cranial nerve connecting receptors in the ear to the brain, plotted according to their best excitatory frequency (the frequency to which they are most sensitive). Amphibian papilla (AP), with a low- and a mid-frequency peak, and basilar papilla (BP) neurons, with a high-frequency peak, are indicated. "Simple units" are neurons that are not influenced by activity in other parts of the ear. "Complex units" are lower-frequency-tuned AP neurons that are suppressed when mid-frequency AP receptors are activated. (*B*) Energy distribution in a synthetic call that stimulated evoked calling at levels equal to a natural call. Note that the peak energy in the stimulus matches the peaks in the lower AP and BP populations of the ear (dashed lines). (*C*) Energy distribution in a natural bullfrog call. Comparing A, B, and C, one can see that the bullfrog ear is matched to the frequency characteristics of the advertisement call used for species recognition and mate choice.

that many cells there respond preferentially to (match) the particular features of a conspecific call and are relatively insensitive to (filter out) general sounds like environmental noise or the specific features of heterospecific calls.

The torus occupies a key position in the neural network processing communication signals. Its output distributes auditory information to lower brainstem motor centers as well as higher motor, motivation, and physiological regulatory areas of the forebrain. Studies by Kim Hoke using immediate early gene (IEG) expression as a marker of neural activity in túngara frogs have shown that the pattern of activity across the torus contains sufficient information to reliably distinguish conspecific from heterospecific calls. Moreover, sex differences in toral activity match sex differences in behavioral selectivity to calls. Male túngara frogs respond by calling back to a much wider range of call stimuli than females respond to with phonotaxis; toral activation is similarly more permissive in males than in females. A complementary IEG study by Kathleen S. Lynch found that hormonal modulation of toral activation by conspecific calls matched hormonal modulation of female behavioral responses to such calls. This midbrain area appears to serve as an important neural processor of calls and regulator of their transformation into a behavioral response. Twice before in this book we have seen something similar—in the jamming avoidance response of electric fish (Chapter 3) and the prey-catching system of frogs (Chapter 4). In these examples, we also see peripheral sensory matching to behaviorally relevant stimulus features, the gradual construction of feature detectors as central nervous system processing proceeds, and a midbrain structure serving as an important interface between sensory input and behavioral response.

Franz Huber played a leading role in establishing neuroethological investigations of cricket acoustic communication that provide an invertebrate perspective complementary to the vertebrate work on frog signal processing. Like frogs, crickets produce acoustic advertisement signals, although they employ a stridulatory mechanism that is independent of their respiratory track. Males produce trains of pulses comprising their familiar chirps by rubbing one body part—a scraper—against another body part—a file. In crickets such as *Teleogryllus oceanicus*, the pulse repetition rate is the key call feature distinguishing the conspecific call from heterospecific calls and the feature that females use for species recognition and mate choice. The receptors in cricket ears (located on their forelimbs, not their heads) contain two (sometimes three) discontinuous populations of receptors (Fig. 7.3). The population of receptors tuned to lower frequencies is most sensitive to audible sounds ~4–5 kHz, whereas the population tuned to higher frequencies is sensitive to ultrasonic sound generally above 20 kHz. Most of the energy in cricket calls matches that lower peak, which then acts like the matched filters proposed by Capranica, allowing acoustic energy in the calls while filtering out frequencies much below and above them.

The key for the cricket, however, is not the call spectral frequencies but the pulse repetition rate of the chirps. Gerald S. Pollack and collaborators Kazuo Imaizumi and Patrick Sabourin found that the activity patterns of individual peripheral receptor cells

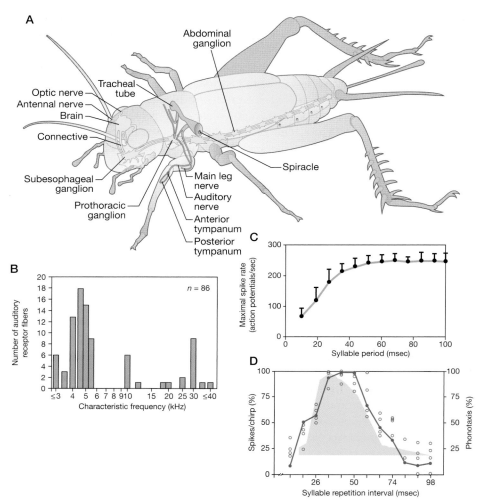

Figure 7.3. (A) Drawing of a cricket showing the location of the ears (anterior and posterior tympanum) just below the forelimb knee and various segments of the nervous systems. Auditory nerve fibers from auditory receptors in the ear connect to the prothoracic ganglion. (B) Distribution of auditory nerve fibers from the ear by their characteristic, or best excitatory, frequency. The low-frequency population is tuned to the energy of the conspecific chirp; the higher-frequency population is tuned to ultrasonic frequencies used by bats, a significant predator of crickets. (C) Responses of auditory fibers (in spikes per section) in the ear to chirps of different repetition rates. Note that the neurons can respond to a large range of chirp rates, but are not "tuned" to a specific rate. (D) Cells in the prothoracic ganglion are more selective to chirp rate and are tuned (respond best) to chirp rates characterizing the conspecific call (red line). Their response pattern matches the cricket's behavioral response to chirps of various rates (indicated as phonotaxis rate, shaded area).

could follow the chirp pattern, but fairly poorly. However, on a statistical level across the low-frequency population, the ear was well able to encode the pulse rate. Peripheral cells are not "tuned" to particular pulse rates, however; their response merely follows whatever they encounter. Franz Huber and his colleagues found that the brain interneurons receiving the inputs from the ear do, however, use that information to construct temporal feature detectors that are tuned to the range of pulse rates that characterize the conspecific signal. They also receive input from each ear and therefore are in a position to calculate the location of incoming sound by comparing interaural inputs. These interneurons in the central ganglia (termed "omega neurons" because of their shape) are the cells that, in turn, connect to motor circuits in the nervous system's lower ganglia responsible for phonotaxis. Despite the differences in nervous system organization, the general processing parallels with frogs are obvious. But what about the other population tuned to ultrasonic frequencies? Unlike the high BP population in frogs, they play no role in mate choice. Instead, these receptors are tuned to the ultrasonic hunting signals emitted by crickets' major predator, bats. When activated, the ultrasonic sensory channel triggers escape behavior rather than approach.

As we have noted before, signal processing for species recognition has implications for intraspecific mate choice preferences. An understanding of the way neurons process signals helps to understand why. Their responses are not all or none, but, rather, are graded around some peak response. Signals falling on the peak elicit the maximum response, whereas those falling progressively above or below elicit proportionally lower responses until the point at which the neuron (and any behavior it triggers) no longer responds. This means that the variation in conspecific signals will translate to variation in the strength of the neural response, even though all of them will elicit some activity. In many frogs, for example, females prefer males with calls that are below the mean frequency for the population. James H. Fox modeled responses of the BP in túngara frogs and was able to show that variation in its activation by the high-frequency "chuck" in the túngara frog call predicted this bias toward lower-frequency male calls. Capranica, in fact, noted in his 1968 paper that the lower-frequency energy in the calls of juvenile bullfrogs generally falls ~700 Hz, within the mid-frequency rather than the low-frequency AP population. As the bullfrogs mature and grow larger, simple biomechanical processes lower the frequency of their calls toward the low-frequency population. This translates into intraspecific behavioral preference: Even though a receiving bullfrog can hear calling bullfrogs of all ages, only the larger, more mature males have low-frequency call components that strongly stimulate the low-frequency AP population needed for call recognition.

LEARNING MATE RECOGNITION

In many, if not most, cases of sexual communication, there seems to be a strong genetic basis to recognize conspecifics. In crickets and frogs, for example, there is no suggestion that learning influences either the calls a male produces upon reaching sexual maturity or the calls to which a female is attracted. There are, however, some spectacular examples of animals learning conspecific recognition, the most popular of which might be filial imprinting. Here there is a sensitive phase when young learn characteristics of their parents that guide their species identification throughout their lives. This phenomenon was popularized by Konrad Lorenz, who showed that greylag geese imprinted on any moving object to which they were exposed 12–16 h after their birth. Some geese imprinted on Lorenz and followed him wherever he went.

Song Learning in Songbirds

The best-studied system of learned mate recognition signals is song learning in songbirds. Peter Marler and his colleagues used Kaspar Hauser experiments, most prominently in white-crowned sparrows, in which males were deprived of song during specific times to uncover the details of song learning (see Fig. 7.4, bottom, for time line of the zebra finch). Birds do not learn just any song but have a genetic bias toward learning certain types of sounds, which they learn by listening to the songs of adult birds. It was thought that males needed to be exposed to song during the first 10–50 d of their life. Later studies, however, showed that some song acquisition occurs later and that learning takes place in two phases, the song acquisition phase and the sensory phase. During the song acquisition phase, young birds listen to songs and memorize them. Young male white-crowned sparrows do not produce sound until some weeks or months later. During the sensorimotor phase when singing starts, there is an initial overproduction period in which males produce a variety of sounds, but this variety is winnowed down until males crystallize some variant of their species-specific song modeled after what they heard in the earlier song acquisition phase. Males must be able to hear their songs when producing them for song development to proceed in an orderly manner. And mistakes are made. Small errors in song learning do not disrupt the mate recognition form of the song, but they are passed down, culturally, to future generations and provide the foundation for new dialects, regional variants of the same song.

There are more than 4000 species of songbirds, and they do not all fit the white-crowned sparrow or, as discussed below, the zebra finch models. As noted by Michael Beecher and Eliot Brenowitz, there is substantial variation among species: (1) when they learn their songs, be it in the first few months of life, over the first year, or throughout their lifetimes; (2) if they learn a single song or a repertoire of songs; (3) if they have repertoires, if their song repertoires develop through imitation, improvisation, or invention; and a suite of other differences. Most studies on song learning have focused on a few model systems, but clearly this is at the risk of missing the details of how song learning can be sculpted to meet the demands of the species' ecology, life history, and mating system.

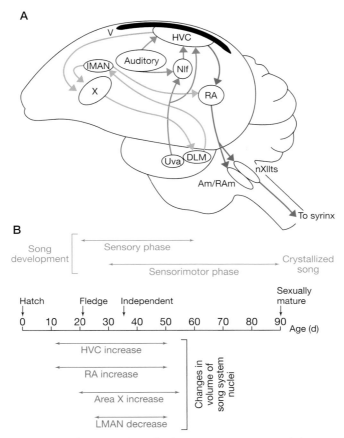

Figure 7.4. The song control system of songbirds. (A) Representations of the neural connections between the major nuclei in the songbirds' song control system. The motor pathway (green) controls the production of song and consists of descending projections from the HVC (high vocal center) in the nidopallium to the robust nucleus of the arcopallium (RA), and from there to the vocal nucleus nXIIts (tracheosyringeal part of the hypoglossal nucleus), the respiratory nucleus etroambigualis (RAm), and the laryngeal nucleus ambiguus (Am) in the medulla. Motor neurons in nXIIts innervate the muscles of the sound-producing organ, the syrinx. (Blue lines) Afferent inputs to HVC from the thalamic nucleus uvaeformis (Uva) and the nidopallial nucleus interface (NIf). (Red lines) Auditory input to NIf and HVC from telencephalic auditory regions. (Orange lines) The anterior forebrain pathway (AFP) that is essential for both perception and learning of song. AFP indirectly connects HVC to RA, through area X (X; thought to be a basal ganglia homolog), the medial portion of the dorsolateral nucleus of the thalamus (DLM), and the lateral portion of the magnocellular nucleus of the anterior nidopallium (IMAN). IMAN also projects to area X. (V) Ventricle. (B) A time line of major life history events in the zebra finch that are related to song learning and the development of the song control system. Unlike song learning in the white-crowned sparrow in which there is no overlap in the sensory and sensorimotor phases, there is substantial overlap of these two phases in the zebra finch. The memorization phase spans ~25–65 d of age, and the motor phase begins at ~30 d of age and continues until crystallized song production. Finches fledge at ~20 d of age and are not fully independent from parental care until ~35 d of age.

The current model most used for the neurobiology of song learning is the zebra finch. This bird is easily kept in captivity, breeds readily, and reaches sexual maturity in a few months. Maturation is so fast that the birds overlap in the sensory phase of song acquisition and the sensorimotor phase of song production. Zebra finches have been important in identifying the changes in the brain that covary with song learning. In all songbirds, the anterior forebrain pathway (AFP; Fig. 7.4) is essential for normal song learning. As Brenowitz and Beecher outline, there are several anatomical changes that take place in zebra finches during the specific time when the birds overlap in the sensory acquisition and sensorimotor phases: in the lateral magnocellular nucleus of the anterior nidopallium (lMAN) shell, axon terminals from neurons of the medial dorsolateral nucleus of the thalamus (DLM) retract; dendritic spine frequencies and the number and density of synapses on lMAN shell neurons decrease; projections from the lMAN core to the robust nucleus of the arcopallium (RA) are remodeled to develop topographic specificity; and new neurons are added in large numbers to area X, which is thought to be a homolog of the basal ganglia. These data are correlative, however, and it is not clear which of these changes in the brain are related to song learning or just a general feature of the rapid development of the zebra finch brain during this time. And if these changes are related to song learning, the overlap of the sensory and sensorimotor phases makes it difficult to ascertain which changes are related to song memory and which are related to song learning.

In addition to neuroanatomical modifications, there are also physiological changes in the AFP during song learning. Long-term potentiation, which enhances synaptic strength and is mediated by activation of NMDA receptors, can be induced by paired stimulation at lMAN synapses before the onset of sensory learning, but the same stimulation produces synaptic depression in birds when the sensory learning phase normally ends. Questions about neural changes underlying song learning might yield to a more comparative approach examining different species with different time lines of song acquisition. In this set of problems and in others in this chapter, we see that model systems are important for elucidating details of a system but usually must be complemented by comparative studies to reach a more general understanding of why animals behave the way they do.

Other Experiential Effects on Mate Recognition

Most studies of songbirds have emphasized learning of the song by males, but there are some data showing that females also learn the song that they will later recognize as indicating a compatible mate. In songbirds in particular, and in animals in general, there are fewer studies of learned mate recognition. A recent, seemingly unlikely, one is a case of sexual imprinting in the cichlid fish.

Species of cichlids in the African Great Lakes can be genetically quite similar but can differ in colors, which, as we reviewed in Chapter 6, can influence female mating recognition. Mate recognition might be learned, although in an unusual way. Female haplochromine cichlids, such as those in the species pair *Pundamilia pundamilia* and *Pundamilia nyererei*, brood their young in their mouths and then continually guard their free-swimming offspring. The offspring are continually exposed to the mother's phenotype during early development, and aspects of her phenotype could act as a model for learning species-specific cues for mate recognition.

Machteld Verzijden and Carel ten Cate conducted cross-fostering experiments in which cichlid offspring were fostered by females of their own species and members of the other species pair (Fig. 7.5). When the female offspring reached maturity, they were then tested for their mate preferences in response to a male of each species. As seen in Figure 7.5, fostered females of both species preferred males of the same species as their foster mother, whether the mother was a conspecific or heterospecific. Although these results clearly show early imprinting that influences later mate preferences, the study raises some important questions. As detailed in Chapter 6, males of closely related cichlid species can differ substantially in color, and color is important in mate choice. It is also known that species recognition is disrupted when color is not reflected by males in the dark, eutrophic waters that occur at some localities and some depths. Unike their male partners, female cichlids are not brightly colored.

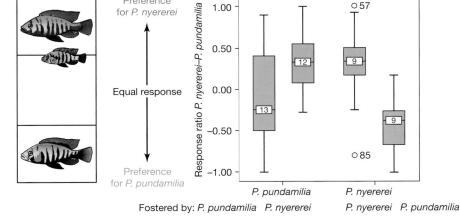

Figure 7.5. Cross-fostering experiments showing maternal imprinting of mate recognition in haplochromine cichlids. (A) The two-way female mate-choice experimental setup. Females can enter male territories through grids, whereas the larger males cannot leave theirs. (B) The results of the cross-fostering experiment. For this figure, data show the approach ratio to *P. pundamilia* subtracted from *P. nyererei*. Scores above 0 reflect a higher approach ratio to *P. nyererei* males, and scores below 0 reflect a higher approach ratio to *P. pundamilia* males. Graphs represent median, interquartile, and full ranges. Numbers on median bars show the number of individuals in each treatment group. (Coloration in the drawings is for cartoon purposes only; the colors shown do not reflect natural coloration.)

They do show some colors that are similar to those of the males, but to a much more subtle extent. Perhaps this cue is enough to influence imprinting. Chemical cues are also known to be important for mate choice in a variety of fishes, and the young might be exposed to strong doses of maternal odors while in her mouth. If these olfactory cues covary with species color, then the mate preferences shown by imprinted young might be mediated by odor and not color.

Learning of mate recognition signals is nearly always restricted to imprinting during early development, but can result from more general experiential effects during other times of an animal's life history. Eileen Hebets exposed female subadult wolf spiders, *Schizocosa uetzi*, to males with a manipulated phenotype; that is, their forelegs were painted either brown or black, which are the two ends of the continuum of natural variation. This experiment considered mating signals that vary *within* rather than *between* species. Upon reaching maturity, females were exposed to courting males of each phenotype. Females were more likely to copulate with males of the familiar phenotype. More strikingly, they were more likely to eat males of the unfamiliar phenotype!

As the studies of cichlids and wolf spiders show, experiential effects of mate recognition and mate choice might lurk in some obscure corners of an animal's biology. As with bird song, we expect there to be genetic as well as experiential influences. A better understanding of how animals acquire a percept of their ideal mate and how the environment interacts with genetic biases is clearly needed.

REPRODUCTIVE SYNCHRONIZATION

Above we noted that one function of sexual communication is to allow individuals to recognize members of their own species and discriminate against those of other species to avoid the deleterious consequences of mating with genetically incompatible individuals. Another function is to synchronize the behaviors of males and females to facilitate the exchange of gametes and coordinate their extended reproductive efforts, such as building nests or provisioning offspring. Environmentally or internally triggered changes in endocrine state help to synchronize the reproductive efforts of males and females. As reviewed in Chapter 6, simultaneous seasonal changes in gonadal steroid hormones (testosterone in males, estrogen and progesterone in females) ready both sexes physiologically to reproduce. They also trigger sex-specific reproductive behaviors. Hormones synchronize the reproductive readiness in males and females, and, as neither can directly ascertain the fertility of the other, the communication signals serve as a proxy and direct male–female efforts to opposite-sex individuals with matching reproductive states. The behavioral interactions also allow a finer-grained synchronization leading to effective copulation. Many birds either duet to establish a pair bond leading to mating, or females exchange copulation solicitation displays (subtle postural and wing position changes) with calling males. In

rodents, olfactory and tactile exchanges coordinate male–female approaches, acceptance, and ultimately the lordosis response in the female, a stereotyped posture that
allows males to mount and copulate with her.

Less obvious is the fact that communication between the sexes can coordinate
the physiological aspects of reproduction along with the behavior. A notable example
of these reproductive interactions has been worked out in elegant detail in the ringdove, initially by Daniel Lehrman and continued by Mae-Feng Cheng (Fig. 7.6).

A male ringdove initiates courtship by producing a "bow coo" display as he struts
toward the female and bows to her. This vocal signal is stimulated by the presence of a
female. Initially the female will avoid the male until he locates himself at a potential
nest site and produces a softer "nest coo." The female will then approach the male.
The male now abandons the nest and coos, struts, and chases after the female as he
produces a bow coo. Once again, the female flees. This ritual repeats itself, and in a
day or two the male does not chase the female but instead allows her to participate in
a nest-cooing duet. The duet lasts for a day or two, during which time the male gath-
ers nesting material while the female continues to coo on her own. Copulation takes
place sometime after the female approaches the male and before she produces solo
coos. Three to five days later, she produces a clutch of eggs. Thereafter, the bonded

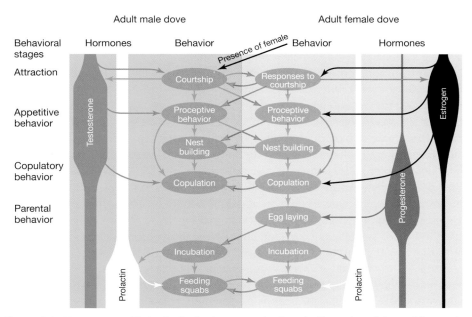

Figure. 7.6. Hormones and behavior in ringdove reproduction. An illustration of the social reproductive interactions between male and female ringdove behavior and the role of their hormones in regulating each.

pair cares for the eggs and later the nestlings. At each stage, the transition from one behavior to the next depends on a behavioral response from the opposite sex.

The hormonal correlates of courting and nesting are well understood in ringdoves (Fig. 7.6; see also Chapter 8 for more information on the endocrinology of male–female interactions). A male's testosterone first increases during the courtship, and his courtship cooing is regulated by this testosterone. Similarly, female estrogen regulates her responses to the male's courtship. It also stimulates maturation of egg follicles and ovulation, and only when those are ready does she allow copulation. Following copulation, the profile of gonadal steroids changes in both sexes (testosterone decreases in males; high estrogen is replaced by high progesterone in females), and in both sexes, the levels of the peptide hormone prolactin rise and stimulate the nesting and offspring care. Lehrman provided a major insight to the field of behavioral endocrinology by showing that these sequential hormonal changes were dependent on the behavioral interactions just as the behavioral sequence itself was; that is, each behavioral step triggered both a behavioral and a hormonal change.

Cheng's work found that this chain of events has some unexpected mechanistic subtleties. It is not the female hearing the male signal that triggers this cascade of behavioral and hormonal changes in the female. Instead, the female must hear herself cooing in response for the endocrine change to occur. Hearing her own responsive coos causes a surge of gonadotropin-releasing hormone (GnRH) from the hypothalamus, which, in turn, results in production of luteinizing hormone (LH) and follicle-stimulating hormone (FSH) from the pituitary, which together promote estrogen production and follicular growth and ovulation. No matter how persistent the male's courtship, the female's follicles remain immature until her vocal self-stimulation. In addition, the female must not only hear the male's coo to initiate this cascade of events, but she must know it is directed toward her. In females that were exposed to bow-cooing males, only females to whom the courtship was directed exhibited significant follicular development. Females can assess whether they are the target of the male's libidinous behavior by watching the direction in which his eyes point. Females not in the male's gaze do not coo in response and do not exhibit follicular development. The female's "state-dependent" reading of the male's intentions might be a mechanism to ensure a pair bond, which would ensure that a female chooses a male committed to the parental duties like nest building that are necessary for successful reproduction. Doves are communal breeders, and obtaining a breeding male should not be sufficient for a female to enter into a reproductive state. Instead, she must know that the breeding male will be committed to breeding with her.

Since Lehrman's classic work, the behavioral interactions of the sexes bringing about both behavioral and physiological reproductive synchrony have become known in other animals. In songbirds, the male's song influences follicular development in the female along with her behavioral responses. In red deer, the bellows of stags stimulate both behavioral and endocrinological receptivity in females. Male calling in túngara frogs triggers approach behavior and elevated estrogen in females. Such effects can also be seen in male–male interactions. Sabrina Burmeister found that exposing male green tree frogs (*Hyla cinerea*) to conspecific calls, which trigger calling in response, resulted in elevations in testosterone. Self-stimulation, as exhibited by the female ringdove, may or may not be part of the mechanism. For example, Burmeister found that the testosterone elevation occurred in males regardless of whether the males hearing the calls responded vocally.

Reproductive synchrony in parthenogenic animals offers an interesting twist to this concept. These are all-female species or strains in which reproduction is clonal and requires no involvement from males. David Crews showed that in a parthenogenic species of whiptail lizard and in a parthenogenic strain of fruit fly, female fecundity is enhanced by the presence of male reproductive behavior. The all-female whiptail lizards (*Cnemidophorus uniparens*) may even alternate between male and female courtship behaviors. Although no gametes are exchanged, and fertile eggs do not depend on the interaction with another individual, individuals will still respond behaviorally to this pseudosexual male behavior, and doing so increases their reproductive success. Thus, even though conspecific males do not exist in these species, females appear to have retained a neuroendocrine axis responsive to male stimulation.

Signal Variation and Physiological Response

As we discussed at the beginning of this chapter, mate recognition signals and preferences for them result in females preferring conspecific signals to heterospecific signals, but also in a preference for some conspecific signals over others. Thus, variation in male mating signals can result in female behavior underlying species recognition, population recognition, and mate preferences within the population. Few studies have asked how signal variation can influence the female's reproductive physiology. Most of these studies have been conducted with songbirds, and the evidence has been mixed.

John Wingfield and colleagues determined the effect of conspecific and heterospecific song on the reproductive state of domestic canaries and wild song sparrows. Both types of song in both types of birds resulted in enhanced follicular development when compared to females who heard no song. Surprisingly, both song types were equally effective in enhancing ovarian development. In canaries there was some evidence of selectivity, because conspecific song resulted in earlier and more frequent oviposition than did heterospecific song.

Scott and Elizabeth MacDougal-Shackleton and colleagues have provided the only study that directly tests whether a female's behavioral mate choice is influenced by intraspecific male signal variation in the same way as her physiological activation. They found that foreign song and natal song of white-crowned sparrows did not differ in their effectiveness in eliciting responses in LHs and ovarian growth when the birds were 1 yr of age. At 2 yr of age, however, females were more attracted to natal call dialects over foreign dialects. Just as for Wingfield's work, there appeared to be dissociation in how the male courtship signals influence female reproductive state and female mate choice. Alternatively, work in canaries suggests that signal complexity may have similar effects in both domains. In many songbirds, including canaries, females prefer more complex songs to less complex ones. In some birds, females supplement their eggs with testosterone. The precise costs and benefits of testosterone enhancement are not known, but it is thought that chicks with a steroid supplement have some competitive advantage in the nest. Canary females that were exposed to more attractive complex songs deposited more testosterone in their eggs than did females exposed to less attractive songs. A similar study did not replicate the same effects on yolk deposition of testosterone, but it did show that the females themselves produced more testosterone if they heard attractive versus unattractive songs.

These few studies do not yet allow us to determine with any certainty whether signals that are more attractive in the sense of being better at soliciting a behavioral response are better at stimulating a physiological response. This remains an important question because understanding the full significance of signal variation is crucial to understanding sexual selection and the evolution of mate recognition systems.

SEXUAL DIFFERENCES IN MATING STRATEGIES, THE FOUNDATION OF SEXUAL SELECTION

In Chapter 2 we noted that Darwin proposed his theory of sexual selection as an addendum to natural selection. Sexual selection could explain the evolution of elaborate traits that are favored by selection, even if they are deleterious to survivorship. The set of traits that drew Darwin's attention all shared some commonalities: They usually characterized males, they were exhibited during breeding, and they seemed to enhance the bearer's ability to do battle with members of the same sex or to be more attractive to members of the opposite sex. These sexually selected traits are usually used by males to attract females, and females assess these traits to choose mates.

Why is it that in most instances of mate choice the males display and the females choose? A classic experiment by Angus J. Bateman on fruit flies provides some important insights (Fig. 7.7). Bateman mated flies from zero to four times and determined the relationship between number of mates and number of offspring. When flies did not mate, they did not reproduce, obviously. Once a female mated, subsequent matings did not influence her number of progeny, whereas males fathered more offspring the more they mated. This simple experiment shows that the sexes, in fruit flies at least, will be under different selection forces to increase their reproductive success and thus their fitness. Males will be under selection to mate often, whereas mating

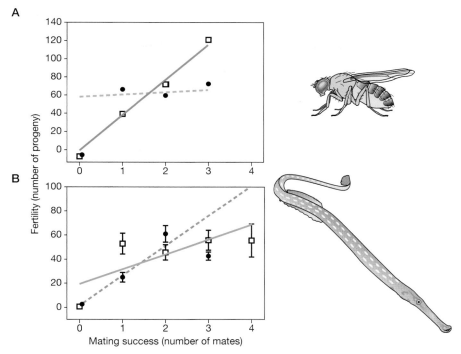

Figure 7.7. Bateman's principle. The effect of the number of matings on male and female fecundity in fruit flies (A) and pipefish (B). In both graphs, the open boxes are males, and the closed circles are females. In both graphs, the solid lines estimate the relationship between mating success and offspring number for males and the dashed lines the same relationship for females.

often, at least in this system, does not increase the females' reproductive success. This simple observation is known as Bateman's principle.

This difference between the sexes arises from difference in parental investment. The most fundamental difference in investment is in gametes. The defining character of sex is gamete size: females have larger gametes, and males have smaller gametes. Typically, once a female commits a gamete to fertilization, she is refractory from reproduction for some time, whereas males are able to continue to reproduce with shorter latency after a mating event. Stated differently, the female has to make a larger investment in producing offspring, which makes her a more limiting reproductive resource than a male would be and consequently gives rise to sexual selection on males. It is more critical for a female to find an appropriate mate, one of the correct species at minimum, than it is for a male because females are more limited in their window for reproduction than are males. A single poor mating, such as with an incompatible hybrid, has a more drastic effect on the female's fitness than on the male's fitness. Thus, selection, in most cases, favors males to mate often and females to mate carefully.

There are many exceptions to Bateman's principle. Göran Arnqvist and Locke Rowe reviewed hundreds of cases in which female insects benefit from multiple matings, in many cases because sperm are bundled in a package that provides females with nutrients or defensive chemicals. In addition, in some cases, the male's investment in offspring is so high that it results in "sex role reversal" in which females compete for males and males choose mates. In Mormon crickets, males put so many resources into their sperm package that they become the limiting sex and females compete for access to males (Fig. 7.8). In another example of sex role reversal, in the fish family Sygnathidae—which includes sea horses, pipefish, and sea dragons—males become pregnant. Females deposit eggs in the male's brood pouch, where he fertilizes and cares for them as they develop. In these species, the male's time commitment to the developing eggs results in his having a larger refractory time after mating than would a female. As we see in Figure 7.7, multiple matings in pipefish have a greater effect on the female's reproductive success than they do on the male's. Although we will concentrate on the more common instance of mate choice being exercised by females, the general principles apply similarly to males choosing females and simultaneous choice by both sexes.

Males and females have different strategies to enhance reproductive success, so it follows that the sexes are under different selection pressures, and, in some cases, these selection pressures can oppose one another. When a result is good for one sex but not for the other, this is called sexual conflict, which is a specific example of a more general phenomenon called antagonistic pleiotropy. As in other cases, fruit flies provide a clear example: Linda Partridge, William Rice, and others have shown that more matings not only enhance fecundity only in males, but that more matings increase female mortality. This is not due to the act of mating per se, but to toxic

Figure 7.8. Courtship provides indirect benefits to females. A female Mormon cricket eating a spermatophylax remaining from a mating. Females receive substantial nutrition from the mass, and this is an example of females accruing a direct benefit that enhances their reproductive success from their mate.

fluids that are part of the male's semen and whose main function is combating sperm of other males. The harm to females is an incidental consequence. It is hardly surprising, then, that males and females can be quite different from one another—so much so that a book on sexual differences in humans by John Gray is entitled *Men Are from Mars, Women Are from Venus: A Classic Guide to Understanding the Opposite Sex.*

It was the pronounced difference between the sexes that cued Darwin to the theory of sexual selection. He noted that males and females differed, by definition, in organs that were critical for reproduction; these are primary sexual characteristics. But males and females are different in a number of other characteristics more peripherally related to reproduction; these are secondary sexual characteristics. Such sexual dimorphisms abound throughout the animal kingdom and result in sexual dimorphisms in body size and coloration, trait adornment and elaborations, and complex courtship rituals. If you open a field guide to neotropical birds to the page on manakins, for only one example, females of different species resemble one another far more than they resemble their own males, and males are quite different from conspecific females as well as from heterospecific males. This is all the result of sexual selection.

BEHAVIORS GENERATING SEXUAL SELECTION

There are two sets of behaviors that give rise to the variation in mating success upon which sexual selection acts. Mate choice between the sexes is one, and competition within the sexes is the other. These are ends of a continuum and are not mutually exclusive within any one mating system. But they offer a helpful heuristic in analyzing the details of how and why sexual selection operates.

Competition between individuals of one sex for access to members of the opposite sex has given rise to a deadly assortment of weapons throughout the animal kingdom (Fig. 7.9). Darwin's suggestion that such weaponry evolved through sexual selection was readily accepted. But Darwin was more interested in the elaborate traits that tended to adorn males despite being a burden for survival (Fig. 7.10). He suggested that these traits evolved under the influence of female mate choice. Darwin did not garner much support from his contemporaries for this idea, in part because he did not offer a cogent explanation for any advantage to the female for exhibiting such preferences, aside from the fact that females have an aesthetic sense for the more adorned males.

Hundreds of studies have now shown that Darwin was correct in his basic prediction that females are more attracted to elaborate traits—that is, to traits that deviate from the population mean in the direction of greater quantity. In the visual domain there are examples of insects, fish, salamanders, and birds with a preference for more courtship behavior; insects, crustaceans, fish, and birds with a preference for

Figure 7.9. Sexual selection favors the evolution of weapons. Examples of weapons in some arthropods and mammals. (A) Weapons in arthropods. Coleoptera. (a) Dor beetles (Geotrupidae). 1, *Lethrus apterus*; 2, *Athyreus nitidus*; 3, *Athyreus tridens*; 4, *Blackbolbus brittoni*; 5, *Blackbolbus lunatus*; 6, *Blackburnium angulicorne*; 7, *Bolborhachium hollowayi*; 8, *Enoplotrupes sharpi*; 9, *Bolborhachium coronatum*; 10, *Lethrus borealis*; 11, *Blackbolbus hoplocephalas*; 12, *Typhaeus typhoeus*. (b) Flower beetles (Cetoniinae). 1, *Cyphonocephalus olivaceus*; 2, *Dicranocephalus bourgoini*; 3, *Eudicella quadrimaculata*; 4, *Ichnestoma rostrata*; 5, *Megalorrhina harrisi*; 6, *Mecynorrhina polyphemus*; 7, *Compsocephalus dmitriewi*; 8, *Theodosia viridiaurata*; 9, *Taurhina polychrous*; 10, *Gnathocera trivittata*; 11, *Anisorrhina algoensis*; 12, *Goliathus albosignathus*; 13, *Mecynorrhina torquata*; 14, *Mecynorrhina passerinii*; 15, *Taurhina longiceps*; 16, *Taurhina splendens*. (c) Dung beetles (Scarabaeinae). 1, *Oxysternon conspicillatum*; 2, *Onthophagus capella*; 3, *Proagoderus rangifer*; 4, *Onthophagus raffrayi*; 5, *Onthophagus dunningi*; 6, *Onthophagus nigriventris*; 7, *Proagoderus lanista*; 8, *Onthophagus mouhoti*; 9, *Onthophagus praecellens*; 10, *Onthophagus sharpi*; 11, *Onthophagus pentacanthus*; 12, *Proagoderus tersidorsis*. (d) Rhinoceros beetles (Dynastinae). 1, *Dynastes hercules*; 2, *Megasoma elephas*; 3, *Eupatorus birmanicus*; 4, *Eupatorus gracilicornis*; 5, *Chalcosoma caucasus*; 6, *Allomyrina* (*Trypoxylus*) *dichotoma*; 7, *Strategus antaeus*; 8, *Enema pan*; 9, *Dipelicus cantori*; 10, *Phileurus truncates*; 11, *Golofa porteri*; 12, *Xylotrupes gideon*. (e) Miscellaneous. (*See facing page for legend.*)

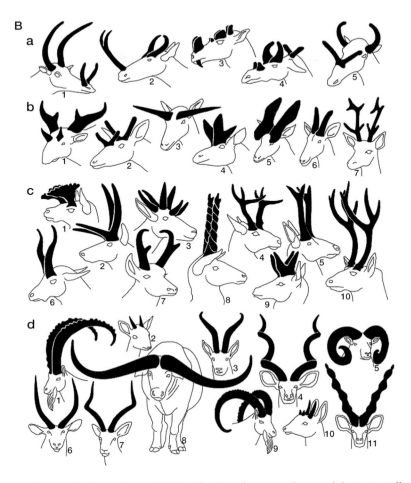

Figure 7.9. (*Continued*). 1, Frog-legged leaf beetle (*Sagra buqueti*; Chrysomelidae); 2, giraffe weevil (*Lasiorhynchus barbicornis*; Curculionidae); 3, baradine weevil (*Parisoschoenus expositus*; Curculionidae); 4, harlequin beetle (*Acrocinus longimanus*; Cerambycidae); 5, rove beetle (*Oxyporus rufus*; Staphylinidae); 6, euchirid beetle (*Euchirus longimanus*; Scarabaeidae); 7, forked fungus beetle (*Bolitotherus cornutus*; Tenebrionidae); 8, *Macrodontia cervicornis* (Cerambycidae); 9, tortoise beetle (*Acromis sparsa*; Chrysomelidae). (*B*) Weapons in mammals. (*a*) Protoceratidae (Tylopoda). 1, *Kyptoceras*; 2, *Synthetoceras*; 3, *Protoceras*; 4, *Paratoceras*; 5, *Syndyoceras*. (*b*) Giraffidae. 1, *Sivatherium*; 2, *Giraffokeryx*; 3, *Canthumeryx*; 4, *Bramatherium*; 5, *Prolibytherium*; 6, *Samotherium*; 7, *Climacoceras*. (*c*) Antilocapridae. 1, *Merriamoceros*; 2, *Tetrameryx*; 3, *Hexameryx simpsoni*; 4, *Paramoceros*; 5, *Paracosoryx*; 6, *Osbornoceros osborni*; 7, *Antilocapra americana*; 8, *Ilingoceros*; 9, *Plioceros*; 10, *Ramoceros*. (*d*) Bovidae. 1, Spanish ibex (*Capra pyrenaica*); 2, dik-dik (*Madoqua kirkii*); 3, Grant's gazelle (*Gazella granti*); 4, kudu (*Tragelaphus strepsiceros*); 5, bighorn sheep (*Ovis canadensis*); 6, waterbuck (*Kobus ellipsiprymnus*); 7, impala (*Aepyceros melampus*); 8, long-horned African buffalo (*Pelorovis antiquus*); 9, Asiatic ibex (*Capra ibex*); 10, chowsingha (*Tetracerus quadricornis*); 11, markhor (*Capra falconeri*).

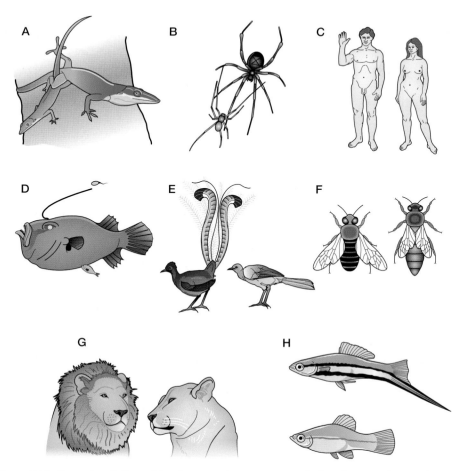

Figure 7.10. Examples of dimorphism between the sexes. (*A*) *Anolis* lizards in copulation. The male (on the *right*), besides being more colorful, has a much deeper and broader head. (*B*) The female black widow spider is substantially larger than her male counterpart, whom she sometimes cannibalizes after sex. (*C*) The basic differences between human male and female anatomy are well known to most readers and are based on the iconic drawing that was on the plaques launched out of the solar system by the Pioneer spacecraft in the hope of making contact with extraterrestrial life. (*D*) The male anglerfish, only a fraction the size of the females, finds a female and then fuses with her for life, and becomes little more than a sperm appendage of the female. (*E*) The male lyrebird has an extravagant tail, which it uses as part of its display when courting the female. (*F*) Male honeybees (*left*), drones, have a morphology, such as large wings, that enhances flight, whereas the reproductive females (*right*), or queens, are designed for reproduction. (*G*) The lion and the lioness are most obviously distinguished by the "kingly" mane of the male. (*H*) Male swordtails possess an elaborated caudal fin in which several rays are extended into a sword-like appendage.

larger body size; and some species in all of these major groups with a preference for males with larger ornaments. There are a plethora of such studies in the acoustic domain as well. Females prefer higher-amplitude calls in insects and frogs; higher calling rates and longer call length in insects, frogs, and birds; more complex signals in frogs and birds; and lower call frequencies in insects, fish, and frogs.

There are fewer examples of elaboration of sexually selected cues in the olfactory domain, but an instructive one involves plant–insect interactions and deceptive mating practices. Orchids are one of the most diverse groups of vascular plants. Like many plants, they rely on insects to transport pollen from one plant to fertilize another plant. Many plants are involved in a mutualism with their animal pollinators, and the plants provide them with rewards, such as nectar, to ensure their involvement in the pollination process. One group of orchids, the deceptive orchids, however, mimics female hymenoptera to dupe males into mating with the flowers. When a male does so, pollen becomes attached to him, which he then inadvertently delivers to another orchid the next time he is wooed by a plant. Deceptive orchids can exploit both olfactory and visual modalities of hymenoptera. As with so many animals trying to attract mates, these plants evolve exaggerated characters, but in this case, the plant's exaggerated characters mimic a female insect. In one case, the orchid pheromones are more potent than those of the female they mimic (Fig. 7.11), and in another, the orchid's flower parts (its labellum) (see Fig. 7.11) that mimic the insect female are longer than those of the model female. The supernormal stimuli in both the odor and visual cues enhance the orchid's attractiveness to its insect pollinator. Not only do preferences for courtship traits influence the evolution of elaborate traits in males through female choice in animal mating systems, but elaborate traits in plants also evolve in response to choice of male insect pollinators.

Returning to animal mating systems, we see that there is tremendous variation in how the sexes are brought together. Males might control territories or other resources to which females need access for reproduction (see Chapter 9). In these systems, we predict that females should choose males with greater resource-holding potential because access to these resources should have a direct effect on female reproductive output. When such is the case, female preferences evolve under direct selection, as defined in Chapter 2. In many insects, such as the Mormon crickets mentioned above (Fig. 7.8), males bundle their sperm in a package—a spermatophylax—which is consumed by the female. Females show preferences for males with a larger spermatophylax, and the spermatophylax can contain nutrients and defensive chemicals that contribute directly to female reproductive success. The males are under selection to provide these nuptial gifts. The female eats the gift while the male mates with her, and the more she has to eat, the longer he is able to mate and the more sperm he transfers, thus ensuring a high level of reproduction.

The mating system that has been the central focus of mate choice, however, is the *lek* (see also Chapter 9). The word derives from the Swedish word for "play," and in mating systems theory, it refers to an arena where males gather to attract females.

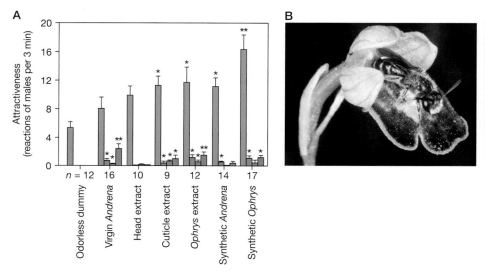

Figure 7.11. Sexual deception in orchids. (A) Reactions of the bee *Andrena nigroaenea* males in behavioral tests in the field. Odorless dummies of female bees scented with different samples of conspecific females (*Andrena*) or orchid flowers (*Ophyrys*) were offered to male bees for 3 min. Means (+5 standard errors) of approaches to the dummy to <5 cm (blue bars), pouncing on the dummy (red bars), alighting on the dummy (green bars), and copulation attempts with the dummy (purple bars) are plotted. Significant differences compared to the same responses to the odorless cue are noted by (*) $p < 0.05$; (**) $p < 0.001$. (B) A male hymenoptera attempting to mate with an orchid. This photo shows an *Ophrys fabrella* and pseudo-copulating *Andrena fabrella* male.

A critical defining characteristic of a lek is that males do not provide resources other than their sperm to females; nor do they later engage in paternal care of the offspring. Thus, there is no opportunity for a female to garner resources from males that might enhance her reproductive success. Yet females show strong preferences when choosing on a lek. As noted above, voluminous data for many species, including lek breeding ones, have demonstrated that females prefer males with more elaborate traits, and it is well accepted that many elaborate traits have evolved under the influence of sexual selection by female mate choice. It is in lek mating systems where the evolution of male traits proceeds to its most extreme. The question is why do females choose if they only receive sperm from their males? There are two issues here that are inextricably intertwined. One issue harkens back to Darwin's notion that females have aesthetic preferences, so we can ask what is it about particular traits that cause them to be reacted to by females as more attractive. The other issue is how does the choice of a certain trait promote the evolution of the female preference for that trait? This, in turn, leads us to ask to what degree does a male's trait indicate aspects about his quality that promote a female's Darwinian fitness?

SIGNAL RELIABILITY

In an ideal "honest" world, aspects of a male's courtship behavior, such as the brightness of his color, the complexity of his song, the strength of his odor, and the vigor of his dance, should all indicate to females his underlying genetic and phenotypic quality. When this is the case, females should evolve to assess a male's courtship accurately, with the result that females choose mates who are better able to provide resources and parental care, or who pass genes to their offspring that enhance survivorship because, for example, they endow their progeny with superior genetic resistance to disease and parasites. If so, all would be in harmony because such a form of eugenics would result in the continuous enhancement of the species' vigor.

The world, however, is neither ideal nor honest. This is true in communication in general, and mate choice specifically, because there is a conflict of interest. The Darwinian interests of the advertising male are not to produce an honest signal but to produce one that will get him mated. The Darwinian interest of the female is to assess the displays in a manner that promotes her overall fitness. An "honest" signal is shorthand for the degree to which a signal reliably indicates some aspect of the sender's quality. Some signals are always fairly reliable. The length of a vertical bar on a fish is a pretty reliable indicator of the fish's size, because the line cannot extend past the fish's body. In other cases it is hard to know when a signal is reliable, especially when signals are supposed to indicate something about the relationship between some of the male's DNA sequences and his ability to survive in the environment. If tail length in a peacock, for instance, is a reliable indicator of heritable variation for survivorship, then what keeps males from "cheating" and producing longer tails? In these cases, either the signals are not reliable or there are forces that make them reliable.

Amotz Zahavi suggested his handicap principle as a resolution to this thorny issue. He predicted that if there is a cost to a signal, which often there is, then only males able to bear the cost will be able to produce the signal. Elaborate male traits are often costly in terms of survivorship, and so only the more physically fit males can exhibit these traits and any "deceivers" would be exposed—not by the females but by predators, parasites, disease, or the normal rigors of everyday life they encounter. Zahavi specifically suggested that such traits honestly indicate genes for enhanced survivorship. Some more recent discussions broaden the definition to include any aspect of the male, his phenotype or genotype, that is indicated by the trait. But this is not what Zahavi was talking about, and it makes the handicap principle so broad as to lose some of its impact.

There is some evidence of honest indicator traits for underlying genetic quality, and some of it comes from peacocks. Peacocks have elaborate trains, or "tails" as they are often called, which are ornamented with large eyespots (Fig. 7.12). The elaborate trains do not enhance survivorship—they probably reduce it. But as in many other animals, females prefer to mate males with these apparently maladaptive traits.

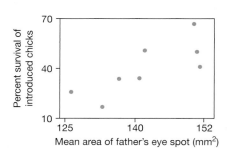

Figure 7.12. Mate choice enhances survivorship in peacocks. Data show that the son's survivorship is correlated with the size of his father's eyespots.

The traits' survivorship costs, however, represent the handicap that Zahavi proposed would keep them honest. This notion was borne out by Marion Petrie's peacock experiments. She randomly mated male peacocks and female peahens, raised their offspring by hand, and later released them into the field. She found that there was a positive correlation between the size of the father's eyespots and his son's survivorship (Fig. 7.12). By preferring certain aspects of the male's phenotype, females produced males that were more likely to survive.

In an intriguing application of the handicap principle to development, Steve Nowicki and colleagues have recently explored how "good genes" associated with developmental canalization can be detected by acoustically based mate preferences in songbirds. As illustrated in Figure 7.4 and discussed above, many areas of the songbird's brain, including nuclei associated with song learning and song production, go through rapid juvenile development. If the animal is under developmental stress, such as starvation, during that time, one assumes that there will be decrements in the growth and development of brain nuclei. Consequently, nutritionally deprived birds might suffer decrements in both the acquisition and production of song, decrements that might be readily apparent to females during mate choice. The evidence offers strong support for this scenario. Summarizing studies of a variety of species, starved birds have smaller telencephalic song control nuclei, specifically, the high vocal center (HVC) and RA (cf. Fig. 7.4), and smaller overall telencephalic volume; they are less accurate in their ability to copy and reproduce their tutor's song; and stressed birds produce songs that are less attractive to females.

There are two extreme evolutionary interpretations to these studies. One is based on incidental consequence. If the probability of a male responding or not responding to this developmental stress has a heritable basis, and there is no evidence either way here, then the female's song-based mate recognition prevents her from hooking up her genes with genes that are susceptible to stress when she discriminates against the songs of compromised males. The other interpretation is based on evolved function. In this scenario, the details of song learning and song production have evolved specifically as handicaps to ensure that only males with superior genotypes, in this case in the context of development, are able to produce attractive song, and that females evolved their song-based mate recognition preferences as a result of a selective advantage accrued by gaining genes that are good for development. Under both interpretations, the functional result is the same: Song variation might be a reliable indicator of early developmental stress, and if there is a heritable basis to an individual's ability to be buffered against this stress, current female preferences result in those genes being recruited for their reproduction. Under both interpretations selection might maintain the song learning–production–recognition system in songbirds. But why did this system evolve—to serve the function of promoting genes for efficient development or for mate advertisement and mate recognition in general? It is not clear at this point, but if the interpretation that the entire song system evolved as a handicap display were true, it would substantially alter our view of the evolution of this most complex communication system.

As noted above, handicaps are not necessary to make all signals reliable. Signal reliability can arise due to design constraints or for reasons that have little to do with communication. Larger male crickets, frogs, and birds have lower-frequency calls because of the biophysical relationship of their sound-producing structures and call frequency. Thus, call frequency is a reliable predictor of size. Another reliable signaling system is genetically based and involves the vertebrate major histocompatibility complex (MHC).

The MHC is a highly polymorphic family of genes inextricably involved in immune function, allowing identification of self versus nonself at the cellular level and thus facilitating defense against invading pathogens. It is one of the most variable genes in the genome because selection favors hypervariability to complement its function in self-recognition. As Wayne Potts and a legion of other researchers have now shown, MHC variation is also correlated with urine odors. Both trained and untrained mice can distinguish odors among MHC variants. Also, in mice, both in the laboratory and in semi-natural conditions, there is disassortative mating by MHC type. Females usually prefer to mate with males whose MHC genotype differs more, rather than less, from their own. The advantage to females of exercising this preference could be derived from enhancing the MHC variability, and thus the immune function, of their offspring. Because MHC is hypervariable, it is also a good marker of kinship. Avoidance of males with similar MHC might engender benefits from inbreeding avoidance. MHC variation has also been shown to influence mate choice in stickleback fish, has been suggested in human mate choice, and modulates kin recognition in tadpoles.

We can conclude that, in many cases, females prefer traits because they are reliable indicators of a male's health and vigor, or because they indicate that he has a genetic constitution that enables such a phenotype. In some cases, such as the peacock's tail, this leads to the evolution of elaborate traits; in other cases, such as MHC-based mate choice, it does not.

SIGNAL DESIGN, SENSORY EXPLOITATION, AND RECEIVER PSYCHOLOGY

There are other alternatives for why females prefer male traits in addition to them being reliable indicators of male quality. One was offered by Ronald Fisher. We reviewed his theory of runaway sexual selection in Chapter 2 in our discussion of indirect selection. Fisher argued that if females prefer a certain male trait, for whatever reason, then a statistical correlation could emerge between that trait and the female preference. The male trait would evolve because it is preferred by females; the female preference would evolve because it "hitchhikes" along with the trait.

Even though Fisher's theory is called runaway, it does not require that traits become more elaborate; they could become less elaborate. It depends on the initial female preference. An indicator trait could initiate the runaway process. In fact, Fisher suggested that females would initially choose males with traits that indicate overall vigor. Thus, the evolution of preferences for reliable traits and the evolution of preferences through runaway sexual selection need not be mutually exclusive.

The design features of any mate recognition systems will inevitably generate sexual selection. As we discussed above, there is strong selection on females to choose males of the correct species for mating, and the design of sensory and neural systems responds to guide her preferentially to those signals. As examples, auditory neurons and photopigments are more likely to respond to certain frequencies of sound or light. Variation is the hallmark of biological systems, so there will always be variation in male mating signals, and some of those mating signals will stimulate the receivers better than others do. This will result in females finding some males more attractive than other males, and to the extent that this translates into some males mating more than other males, it will give rise to sexual selection.

On the one hand, we would expect selection to favor females whose receiving systems evolve to guide her to the males that will most benefit her Darwinian fitness. Even if Zahavi's handicap principle operates in full force, however, there is still a huge stochastic component. Predation will keep less vigorous males from surviving with a longer tail on average, but there will still be substantial variation in how reliably a long tail indicates good genes versus good luck. On the other hand, however, we must remember that eyes and ears and nares did not originally evolve for mate choice, nor is mate choice their only current function. Consider photopigment sensitivity in fish. Fish use their eyes to forage as well as choose mates. A change in sensitivity would have pleiotropic effects because it would influence a female's foraging efficiency as well as which male nuptial colors she finds attractive. There are various outcomes of pleiotropy, but Molly E. Cummings has shown convincingly in surfperches that the fishes evolve photopigment sensitivities that enhance their ability to detect their primary prey against the particular ambient light environment in which they live, and males, in turn, evolve colorations that match those sensitivities. This is called sensory exploitation when males evolve traits to match preexisting female biases and is an alternative scenario to traits and preferences always evolving in a lockstep coevolutionary fashion. Sensory exploitation also occurs in cuckoos, which exploit the perceptual systems that their foster parents evolved to feed their own offspring, and the deceptive sexual signaling used by orchids to dupe insect pollinators. In both cases, one would expect the birds and the bees to evolve a recognition mechanism that would allow them to discriminate a cuckoo nestling from their own offspring, and a plant from a female insect—but they do not. This is either because they cannot, or because the trade-offs in costs and benefits favor retaining the recognition mechanisms they have in place.

In Chapter 1 we reviewed evidence that male swordtails evolved their swords to exploit preexisting biases in females (see Fig. 1.3). There are other examples of sensory exploitation. One occurs in the "spermatophylax-eating" behavior of some insects, discussed above, which also seems to have evolved by sensory exploitation (see Fig. 7.8). Scott Sakaluk has suggested that these nuptial gifts evolved as a form of sensory trap that exploits the normal gustatory responses of females and thus favors the evolution of gift-giving in males, who themselves gain an advantage because they are able to mate more while females are eating. He showed that female crickets of non-gift-giving species, which is the ancestral state of this behavior, provisioned with novel food gifts were "fooled" into accepting more sperm from their mates.

Another example occurs in studies of túngara frogs by Walter Wilczynski, A. Stanley Rand, Michael J. Ryan, and colleagues. As we reviewed in the section on mechanisms of mate recognition, frogs have two inner ear organs, the AP and BP, whose tuning coincides with the spectrum of the male's mating call. In this species, the tuning of the AP matches the spectrum of the recognition component of the call, the whine; and the BP matches the adornment to the call, the chuck. Most of the close relatives of túngara frogs only have calls with whines and lack adornments that stimulate the BP. Nevertheless, most of these species have BPs with identical tuning. Phylogenetic reconstruction shows that BP tuning in túngara frogs is the ancestral condition. It appears that the frequency aspects of the chuck evolved to match the preexisting tuning of the BP as opposed to the alternative that BP tuning evolved to match the particular chucks of better males.

Biases that can be exploited by males are not restricted to the tuning of peripheral end organs, such as eyes and ears. Sexual selection is responsible for increased elaboration and complexity in signals. One prime example is the complex song repertoire of many birds. Many studies have shown that female birds prefer males with larger repertoires. This might be because these males have better genes for survivorship. Complex song might also be more attractive to females because of its influence on habituation. From our own experience, we know that we habituate more rapidly to a simple repeating acoustic stimulus, such as a metronome, than we do to a varied and intricate series of sounds, such as an innovative jazz performance by Charlie Parker.

Almost half a century ago, Charles Hartshorne, a noted Kantian philosopher and avid birder, wrote a paper and later (1973) a book, *Born to Sing: An Interpretation and World Survey of Bird Song*, in which he suggested that complex repertoires in birds evolved to release listeners from habituation. Recent studies, reviewed by David Clayton and colleagues, support this idea. Songbirds often respond actively to an initial stimulus, but that response degrades with repeated presentations of the stimulus. Female grackles, for example, perform courtship solicitation displays to initial songs. The females stop responding if the same song type is repeated but respond again when the presentation is switched to new song types. Neurophysiological and genomic studies of the zebra finch show the same patterns of response at finer levels of organization. Auditory neurons in the caudal medial nidopallium, a part of the telencephalon involved in song production, always show decreased spike rate when one syllable of a zebra finch song is followed by that same syllable. When that same syllable is followed by a different syllable, spike rates again increase. The release of habituation effect in response to new syllables is also apparent at the genomic level. Expression of an immediate early gene, ZENK, is higher in this area of the brain of birds after exposure to syllable 1 followed by syllable 2, compared to those exposed to syllable 1 followed by syllable 1. It appears that in songbirds, and possibly in other taxa, complex signals are better at maintaining the attention of a receiver, and that reason, alone or in part, might be why complexity is ubiquitously preferred by females.

In Chapter 6 we also discussed a study of mate choice in túngara frogs that offers an additional mechanism by which signal complexity might be favored by sexual selection. Females were more likely to remember mating calls with more rather than fewer components. Again, the premium placed on signal complexity by females of so many species in all sensory modalities might have a general explanation as foreshadowed by Darwin:

> When animals utter sound in order to please the females, they would naturally employ those which are sweet to the ears of the species; and it appears that the same sounds are often pleasing to widely different animals, owing to the similarity of their nervous systems.
>
> —Darwin 1872, p. 91

When males evolve signals to match preexisting biases, the end result still can be advantageous for females. The match between the male's signal properties and the female's sensitivities should make the signals more conspicuous and thus reduce the female's costs of searching for them. This was shown quite convincingly by

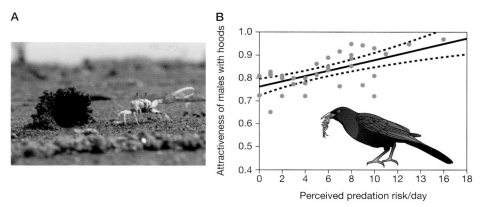

Figure 7.13. The attractiveness of male fiddler crabs with hoods depends on the level of perceived predation risk. (A) A male fiddler crab, *Uca terpsichores*, displaying in front of its sand hood. The sand hood serves as a sexual signal; it projects above the visual horizon of the crabs and marks the entrance to the male's burrow. For scale, males are ~1 cm wide. (B) Grackles eat these crabs. Perceived predation risk is estimated by the number of times females retreated to their burrows when grackles were present. The attractiveness of males with hoods increased significantly with an increasing level of perceived predation risk ($n = 34$, $r^2 = 0.379$, $t = 4.418$, $p < 0.0001$, excluding two outliers; 95% confidence limits shown).

John Christy and his colleagues in a study of fiddler crabs. In many of these species, a male constructs a tower or hood from sand that marks the male's burrow entrance and enhances the contrast of the male's claw-waving courtship display (Fig. 7.13). These ornaments are especially effective at being detected by the crabs, which have evolved visual systems adapted for detecting objects that project from the horizontal plane. In general, males displaying near these structures are more likely to be chosen by females than are males without structures. In one species of fiddler crab, these researchers hypothesized that if towers reduce the search costs for females and offer them a safe haven, then the attractiveness of these ornaments should be enhanced when there is greater predation risk. This is exactly what they found: Males with ornaments were more likely to be chosen by females when there was heightened predation risk by crab-eating birds (Fig. 7.13). Thus the perception, detection, and preference of elaborate and conspicuous courtship can result in a survival fitness benefit to the female that is unrelated to anything about the quality of the male. The females' biases are exploited, but not at a cost to her Darwinian fitness.

The emphasis of research on female preferences that cause trait elaboration in males has focused on genetic benefits via the handicap principle. Other forces, however, such as runway sexual selection and sensory exploitation offer alternative explanations that can either work alone or interact with each other and with indicator traits to explain why females evolve preferences for male traits. We also see that an understanding of female preferences is advanced with an understanding of the mechanisms that underlie them and the phylogenetic context in which they arise.

MATING SYSTEM GENOMICS

There have been two goals in studying the genetics of male traits and the preferences for them. One deals with ultimate questions and the other with proximate questions.

The first goal is concerned with the potential for evolution of male traits and female preferences and is conducted in the realm of quantitative genetics. There is variation in male traits, but traits will only evolve under sexual selection if a portion of the overall phenotypic variation is due to genetic variation (i.e., if the traits are heritable). There are some studies that show this can be the case. In sticklebacks and guppies, females prefer males with more nuptial coloration, and there is a significant correlation between the amount of nuptial coloration between fathers and sons. Taking another approach, David Hosken and his colleagues showed that attractiveness itself can be heritable. In the fruit fly *Drosophila simulans*, they measured the attractiveness of fathers and their sons who were generated by a full-sib/half-sib mating design. There was significant heritability of attractiveness through sires ($h^2 = 0.29$), but not through dams ($h^2 = 0.14$). It was not clear which male traits were responsible for the male's attractiveness, but even when there is a single trait that explains female mating preferences, attractiveness is likely to be a composite trait, with some traits having more or less influence on females, so studies of heritability of attractiveness, such as this one, seem especially germane.

Breeding experiments have also revealed that there can be substantial heritability in female mating preferences. As reviewed by Michael Jennions and Marion Petrie, studies have revealed significant genetic variation for female mating preferences in guppies, stalk-eyed flies, bollworms, planthoppers, and grasshoppers. In these studies, either the mating preference was directly selected for or the preferred trait was selected for and the preference changed as a correlated response. Thus, there can be substantial genetic variation in both male traits and female preferences. These are evolvable traits that have the genetic capacity to respond to selection once it occurs.

The second goal of genetic studies of mating systems is to identify the genetic mechanisms that result in male traits and female preferences. Here the evidence is quite scant. The best, most convincing evidence comes from studies of the *period* gene in *Drosophila*, which we reviewed in Chapter 3. Besides controlling circadian rhythms, this gene also controls the period of the male's love song, an ideal example of pleiotropy. As few as four to five amino acid substitutions of the gene's longest exon explain the difference in these mate recognition traits between closely related species.

One system that has emerged as a promising link between genes, behavior, and evolution is the swordtails of the genus *Xiphophorus*. In two species of swordtails, *Xiphophorus nigrensis* and *Xiphophorus multilineatus*, there is a polymorphism in male body size. A series of breeding experiments with the congeneric platyfish by Klaus Kallman and his colleagues beginning in 1978 suggested that this variation was under control of a single Y-linked locus called the *P* gene for its hypothesized effect on the male's hypothalamic–pituitary–gonadal (HPG) axis. It appeared that Y-linked allelic variation is responsible for initiating puberty at different times and ages in males. As with all live-bearing fishes, males cease or drastically reduce growth when they reach sexual maturity. The result is that some genotypes mature soon and small, and others mature late and large. The heritability for body size is >0.90 in *X. nigrensis*.

X. nigrensis has three size classes of males: small, intermediate, and large. In a series of studies of their mating system, summarized by Michael J. Ryan and Gil G. Rosenthal, it was shown that the different male phenotypes exhibited different mating behaviors. Large males court females with a series of showy behaviors that expose both their dorsal fins and swords, whereas small males chase females and try to force copulations. Most intermediate-sized males court, but the smaller males of that size class can also chase. Because the size variation is Y-linked, the strength of sexual selection in the wild on male size and the alleles determining it can be ascertained from paternity analysis by comparing the size distributions of fathers and their sons, the latter of which are raised in the laboratory from field-collected gravid females. Large and intermediate males are overrepresented in the F_1 generations relative to their proportions in the P generation. Thus, in the wild, sexual selection favors larger males over smaller males. Laboratory studies show that the sexual selection advantage is likely due to a strong female preference for larger males. If larger males are more successful, what maintains the genetic variation for small size? Although small males have a disadvantage in mate choice, they have an advantage in reaching early maturity because they can then mate at a younger age. Population models show that the system is in equilibrium: The mating advantage to large males is offset by the advantage of small males in surviving to maturity.

The long-sought *P* gene hypothesized by Kallman was recently mapped, sequenced, and analyzed by Manfred Schartl and his colleagues. The gene was mapped to near the sex-determining locus on the X and Y chromosomes. In this region there are several copies of the melanocortin 4 receptor gene (*mc4r*). *mc4r* is a seven-transmembrane G-protein-coupled receptor, expressed in high concentrations in the hypothalamus and linked to regulation of the energy budget through its regulation of food intake. It has also been proposed to interact with the hypothalamic–pituitary–thyroid axis. A link to the HPG axis seems likely as well, because *mc4r* influences leptin signaling leading to GnRH production.

mc4r allelic variation falls into three classes—A, B1, and B2—which differ in the carboxy-terminal region of the receptor protein, shown in Figure 7.14. B alleles differ from A alleles in the lack of cysteine residuals in the carboxyl terminus. A alleles are present on the X chromosome, and both types of B alleles are on the Y chromosome. Interestingly, the number of B alleles on the Y chromosome is variable and can range from a few (three and more) up to 10. Male body size did not correlate with specific alleles but did vary predictably with copy number. In addition, large and

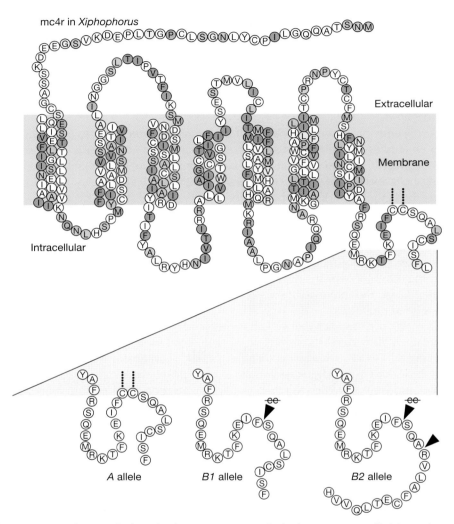

Figure 7.14. In the swordtail, *Xiphophorus nigrensis*, male body size is controlled by a dosage-dependent response to nonfunctional (*B* alleles) *mc4r* genes. (*Top*) Amino acid sequence and schematic structure of mc4r (*A* allele group), which codes for a G-coupled protein receptor (seven transmembrane). The position of each amino acid sequence inside and outside the cell is shown. Colors represent levels of selection (Tajima's *D*) as determined with the program Selecton: (purple) strong purifying selection; (light blue) purifying selection; (yellow) positive selection; (orange) strong positive selection. (*Bottom*) Carboxy-terminal tail of the mc4r. The *A* allele types represent the wild-type sequence. The *B* alleles are characterized by a six-base deletion at the carboxy-terminal end of the protein. Owing to this deletion, all *B* allele types lack two cysteine amino acid residues (CC), which renders them nonfunctional. The *B2* alleles have an additional four-base deletion that leads to a frameshift and elongated protein (seven additional amino acids).

intermediate-sized males had about 10-fold higher expression of *mc4r* in the brain than did small males and females. The most obvious difference between the *A* and *B* allele groups is the presence or absence of the CC motif (Fig. 7.14), which is highly conserved in vertebrate *mc4r* genes. In vitro experiments using cells that stably expressed the *A* or *B* forms of *X. nigrensis mc4r* receptors encoded by both *B* alleles were not stimulated by ligands, whereas receptors with *A* alleles responded with higher cyclic adenosine monophosphate (cAMP) production and reporter gene expression when they were stimulated. Thus, the *A* but not the *B* alleles are functional. The researchers suggest that non-signal-transducing *mc4r* variants (encoded by *B* alleles) reduce the formation of functional *mc4r* receptor dimers or sequester the ligand and therefore delay the onset of puberty. The number of copies of *B* alleles is correlated to male body size; thus it appears that *B* alleles delay the onset of puberty by diminishing the signal from the functional *A* alleles. The more *B* copies present, the longer it takes until a threshold of functional *mc4r* signal is reached that is sufficient for the HPG axis to be up-regulated and sexual maturity to be initiated. In this case, variation in copies of a single gene seems to explain the polymorphism in body size and correlates with a suite of other characters involved in the mating system of swordtails, and the manner in which this happens is understood at the cellular and molecular levels.

CONCLUSION

Mate choice is a critical aspect of an animal's behavior because it determines with which individual it will partner its genes for time travel across generations. Studies of species recognition, the neural mechanisms underlying recognition, and the hormonal synchronization of the sexes provide an integrative framework for understanding this most important social behavior. The studies of species recognition also lend themselves to examination of the variation of traits and preferences that give rise to sexual selection. This selection force is responsible for some of the most diverse phenotypes in the animal kingdom. It is in this context that it becomes especially clear that understanding the evolution of mating signals requires a detailed understanding of the mechanisms of mate preference, including ways in which they might be exploited and the phylogenetic context in which they evolve.

BIBLIOGRAPHY

Arnqvist G, Rowe L. 2005. *Sexual conflict*. Princeton University Press, Princeton, NJ.

Bateman AJ. 1948. Intrasexual selection in *Drosophila*. *Heredity* **2:** 349–368.

Beecher MD, Brenowitz EA. 2005. Functional aspects of song learning in the songbirds. *Trends Ecol Evol* **20:** 143–149.

Bentley GE, Wingfield JC, Morton ML, Ball GF. 2000. Stimulatory effects on the reproductive axis in female songbirds by conspecific heterospecific male song. *Horm Behav* **37:** 179–189.

Brenowitz EA, Beecher MD. 2005. Song learning in birds: Diversity and plasticity, opportunities and challenges. *Trends Neurosci* **28:** 127–132.

Brown WL, Wilson EO. 1956. Character displacement. *Syst Zool* **5:** 49–65.

Cheng MF. 1992. For whom does the female dove coo? A case for the role of self-stimulation. *Anim Behav* **43:** 1035–1044.

Coyne JA, Orr HA. 2004. *Speciation*. Sinauer, Sunderland, MA.

Crews D. 1987. Courtship in unisexual lizards: A model for brain evolution. *Sci Am* **255:** 116–121.

Crews D, Teramoto LT, Carson HL. 1985. Behavioral facilitation of reproduction in a sexual and parthenogenetic *Drosophila*. *Science* **227:** 77–78.

Cummings ME. 2007. Sensory trade-offs predict signal divergence in surfperch. *Evolution* **61:** 530–545.

Darwin C. 1872. *The expression of the emotions in man and animals*. J. Murray, London.

Davies NB, Kilner RM, Noble DG. 1998. Nestling cuckoos, *Cuculus canorus*, exploit hosts with begging calls that mimic a brood. *Proc R Soc Lond B* **265:** 673–678.

Dong S, Clayton D. 2009. Habituation in songbirds. *Neurobiol Learn Mem* **92:** 183–188.

Emlen DJ. 2008. The evolution of animal weapons. *Ann Rev Ecol Evol Syst* **39:** 387–413.

Frishkopf LS, Capranica RR, Goldstein MH Jr. 1968. Neural coding in the bullfrog's auditory system—A teleological approach. *Proc IEEE* **56:** 969–980.

Gill DE, Leboucher G, Lacroix A, Cur R, Kreutzer M. 2004. Female canaries produce eggs with greater amounts of testosterone when exposed to preferred male song. *Horm Behav* **45:** 64–70.

Hartshorne C. 1956. The monotony-threshold in singing birds. *Auk* **73:** 176–192.

Hartshorne C. 1973. *Born to sing: An interpretation and world survey of bird song*. Indiana University Press, Bloomington, IN.

Hebets EA. 2003. Subadult experience influences adult mate choice in an arthropod: Exposed female wolf spiders prefer males of a familiar phenotype. *Proc Natl Acad Sci* **100:** 13390–13395.

Hoke KL, Burmeister SS, Fernald RD, Rand AS, Ryan MJ, Wilczynski W. 2004. Functional mapping of the auditory midbrain during mate call reception. *J Neurosci* **24:** 11264–11272.

Hoke KL, Ryan MJ, Wilczynski W. 2008. Candidate neural locus for sex differences in reproductive decisions. *Biol Lett* **4:** 518–521.

Huber F. 1990. Cricket neuroethology: Neuronal basis of intraspecific acoustic communication. *Adv Study Behav* **19:** 299–356.

Huber F. 2006. Experiences and highlights during my time in Seewiesen 1973–1993. *International Society for Neuroethology Newsletter* (November 2006), pp. 3–6.

Huber F, Thorson J. 1985. Cricket auditory communication. *Sci Am* **253:** 47–54.

Imaizumi K, Pollack GS. 1999. Neural coding of sound frequency by cricket auditory receptors. *J Neurosci* **19:** 1508–1516.

Jennions MD, Petrie M. 1997. Variation in mate choice and mating preferences: A review of causes and consequences. *Biol Rev Camb Philos Soc* **72:** 283–327.

Jones AG, Rosenqvist G, Berglund A, Arnold SJ, Avise JC. 2000. The Bateman gradient and the cause of sexual selection in a sex-role-reversed pipefish. *Proc R Soc Lond B* **267:** 677–680.

Kallman KD, Borkoski V. 1978. A sex-linked gene controlling the onset of sexual maturity in female and male platyfish (*Xiphophorus maculatus*), fecundity in females and adult size in males. *Genetics* **89:** 79–119.

Kim TW, Christy JH, Dennenmoser S, Choe JC. 2008. The strength of a female mate preference increases with predation risk. *Proc R Soc Lond B* **276:** 775–780.

Lampert KP, Schmidt C, Fischer P, Volff J-N, Hoffmann C, Muck J, Lohse MJ, Ryan MJ, Manfred Schartl M. 2010. Determination of onset of sexual maturation and mating behavior by melanocortin receptor 4 polymorphisms. *Curr Biol* **20:** 1729–1734.

Lehrman DS. 1965. Interaction between internal and external environments in the regulation of the reproductive cycle of the ring dove. In *Sex and behavior* (ed. Beach FA), pp. 344–380. Wiley, New York.

Lynch KS, Wilczynski W. 2008. Reproductive hormones modify reception of species-typical communication signals in a female anuran. *Brain Behav Evol* **71:** 143–150.

MacDougall-Shackleton MA, MacDougall-Shackleton EA, Hahn TP. 2001. Physiological and behavioural responses of female mountain white-crowned sparrows to natal- and foreign-dialect songs. *Can J Zool* **79:** 325–333.

Marler P. 1997. Three models of song learning: Evidence from behavior. *J Neurobiol* **33:** 501–516.

Mayr E. 1982. *The growth of biological thought*. Belknap Press, Cambridge, MA.

Moriarity Lemon E. 2009. Diversification of conspecific signals in sympatry: Geographic overlap drives multidimensional reproductive character displacement in frogs. *Evolution* **63:** 1155–1170.

Nabatiyan A, Poulet JFA, de Polavieja GG, Hedwig B. 2003. Temporal pattern recognition based on instantaneous spike rate coding in a simple auditory system. *J Neurophysiol* **90:** 2484–2493.

Nowicki S, Peters S, Podos J. 1998. Song learning, early nutrition and sexual selection in songbirds. *Am Zool* **38:** 179–190.

Partridge L, Farquhar M. 1981. Sexual activity reduces lifespan of male fruit flies. *Nature* **294:** 580–582.

Petrie M. 1994. Improved growth and survival of offspring of peacocks with more elaborate trains. *Nature* **371:** 598–599.

Potts WK, Manning CJ, Wakeland EK. 1991. Mating patterns in seminatural populations of mice influenced by MHC genotype. *Nature* **352:** 619–621.

Reusch TBH, Häberli MA, Aeschilmann PB, Milinski M. 2001. Female sticklebacks count alleles in a strategy of sexual selection explaining MHC polymorphism. *Nature* **414:** 300–302.

Rice WR. 1992. Sexually antagonistic genes: Experimental evidence. *Science* **256:** 1436–1439.

Ryan MJ, Rosenthal GG. 2001. Variation and selection in swordtails. In *Model systems in behavioral ecology* (ed. Dugatkin LA), pp. 133–148. Princeton University Press, Princeton, NJ.

Ryan MJ, Fox JH, Wilczynski W, Rand AS. 1990. Sexual selection for sensory exploitation in the frog *Physalaemus pustulosus*. *Nature* **343:** 66–67.

Sabourin P, Pollack GS. 2009. Behaviorally relevant burst coding in primary sensory neurons. *J Neurophysiol* **102:** 1086–1091.

Sakaluk SK. 2000. Sensory exploitation as an evolutionary origin to nuptial food gifts in insects. *Proc R Soc Lond B* **267:** 339–343.

Schaefer HM, Ruxton GD. 2009. Deception in plants: Mimicry or perceptual exploitation. *Trends Ecol Evol* **24:** 676–685.

Schiestl FP, Ayasse M, Paulus HF, Löfstedt C, Hansson BS, Ibarra F, Francke W. 1999. Orchid pollination by sexual swindle. *Nature* **399:** 421. doi: 10.1038/20829.

Taylor ML, Wedell N, Hosken DJ. 2007. The heritability of attractiveness. *Curr Biol* **17:** 959–960.

Trivers RL. 1972. Parental investment and sexual selection. In *Sexual selection and the descent of man* (ed. Campbell B), pp. 136–179. Aldine, Chicago, IL.

Verzijden MN, ten Cate C. 2007. Early learning influences species assortative mating preferences in Lake Victoria cichlid fish. *Biol Lett* **3:** 134–136.

Villinger J, Waldman B. 2008. Self-referent MHC type matching in frog tadpoles. *Proc R Soc Lond B* **275:** 1225–1230.

Wedekind C, Seebeck T, Bettens F, Paepke AJ. 1995. MHC-dependent mate preferences in humans. *Proc R Soc Lond B* **260:** 245–249.

Wilczynski W, Rand AS, Ryan MJ. 2001. Evolution of calls and auditory tuning in the *Physalaemus pustulosus* species group. *Brain Behav Evol* **58:** 137–151.

Zahavi A, Zahavi A. 1997. *The handicap principle: A missing piece of Darwin's puzzle.* Oxford University Press, Oxford.

CHAPTER 8

Social Bonding and Cooperation

FOR MOST ORGANISMS, REPRODUCTIVE SOCIAL interactions end at mating. But for many animals, mating is just the beginning of an extended period of social relationships. These extended relationships are anchored in three types of social bonding: male–female interactions centered on the care of young, parent–offspring interactions, and interactions among siblings and other close relatives.

MALE–FEMALE BONDING

Mate choice and mating do not always culminate in male–female social bonding or any prolonged social interactions past the copulatory act. However, when they do, there are noticeable asymmetries between males and females in the expression of this behavior.

Monogamy and Promiscuity

Mating systems differ along two dimensions—number of partners and longevity of association. At one extreme is promiscuous mating. Males and females mate and then separate with no further social interactions, and mate with multiple individuals as available and physiologically feasible. Given the different parental investment in gametes (see Chapter 7), males are capable of mating with more females than females are with males. This is especially true when females must also gestate, brood, nurse, or otherwise care for offspring. The discrepancy this generates is most obvious in birds and mammals. On the other extreme is monogamy, when males and females bond during mating and remain a social couple. The male–female bond can last through a single breeding season, but in some cases can endure the life span of the partners. Long-term monogamous social bonding is most common in birds, but is also seen in

---

some mammals. Beyond those taxa, it is rare. Monogamous pair bonding is the result of recruiting male parental care to the maternal care efforts.

Many small rodents demonstrate a promiscuous mating system in which males mate with females whenever the opportunity arises, then abandon the female and take no role in caring for the offspring. Montane and meadow voles, species of the genus *Microtus*, are examples. Males do not pair bond and, in fact, appear indifferent to social interactions other than mating. They show no affiliative behavior toward pups in general and do not contribute to the care of their own pups. Two closely related *Microtus* species, prairie and pine voles, are monogamous. Males form a lifelong pair bond with a female after mating with her. They show a strong attachment to their partners, and, in laboratory tests, they strongly prefer to be close to her over other females. Monogamous vole males then care for offspring, taking turns with the female guarding the nest. They are selectively aggressive toward unfamiliar individuals, indicating that their behavioral tolerance of their partner and offspring is not the result of being generally docile. The promiscuous species show limited, and undiscriminating, investigatory approaches to both familiar and unfamiliar females and respond to social odors no differently than to other odors. Monogamous prairie voles instead approach novel social odors more quickly and spend more time investigating them than they do other odors. The species do not differ in their investigation of novel nonsocial odors such as food odors or arbitrary chemicals, or in their ability to discriminate them. The species differences are restricted to the males. Female behavior is essentially identical among the species.

Larry Young has exploited variation in mating systems among vole species to link behavioral, neurobiological, and genetic variation into an integrated understanding of social behavior (Fig. 8.1). Given that the essential species difference is in male behavior, Young and coworkers focused on variation in brain vasopressin systems (see Chapter 3) as a potential mediator of the differences. Treating prairie voles with vasopressin increases the suite of attachment and paternal care behaviors typifying their monogamous lifestyle. Curiously, treating promiscuous meadow voles with vasopressin has little effect. This points to a difference in brain vasopressin receptors, rather than vasopressin levels, as a key factor in the behavioral differences. In fact, the distribution and density of one vasopressin receptor, the V1a receptor, is significantly different in the monogamous and promiscuous species. The density of V1a receptors in promiscuous meadow voles can be artificially increased by using an adeno-associated viral vector (AAV) to transfer the V1a receptor gene of the prairie vole to the meadow vole. The recipient meadow voles then exhibited behavior similar to that of the monogamous voles. They formed pair bonds and expressed paternal care. The conclusion is that both types of voles have brain vasopressin, but the monogamous and promiscuous species differ in the receptor's ability to use it to modulate behavior.

The genetic variation behind these differences is not in the coding region for the receptor itself (which is highly conserved across all vertebrates), but in a regulatory microsatellite region upstream from the 5′ region of the gene. The monogamous

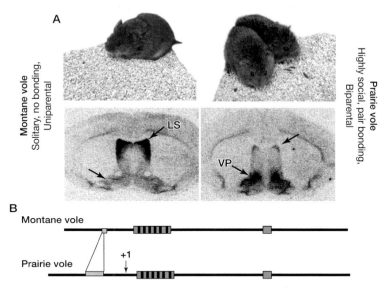

Figure 8.1. (A) Montane voles (*left*) have solitary males and uniparental care; prairie voles (*right*) are social, pair bond, and have biparental care. Vasopressin binding density patterns in the limbic system differ significantly in the two species. LS, lateral septal nucleus; VP, ventral pallidum. (B) Biparental, pair-bonding species differ in a microsatellite region of the V1A receptor gene.

prairie and pine voles have a long version of the microsatellite, whereas the promiscuous mountain and meadow voles have a short version. The long version results from repeats in the 3′-end of the microsatellite. In addition to the species difference, the 3′-end is polymorphic among prairie vole males, resulting in individual differences in microsatellite length. This genetic polymorphism correlates with within-species behavioral variation in attachment: the longer the microsatellite, the greater the social approaches, social order investigation, and strength of partner preference of the male (Fig. 8.2). Young's work shows how a small variation in a gene regulatory sequence leads to a significant variation in the receptors determining how a brain responds to a neuromodulator, which finally leads to profound inter- and intraspecies differences in social behavior. The effects of V1a receptor variation are not peculiar to voles. Hasse Walum and colleagues found that V1a receptor polymorphisms correlate in the predictable way with ratings on a partner bonding scale in human men.

Polygyny and Polyandry

Multiple mates combined with some type of postmating association characterize two other mating systems—polygyny (one male, many females; a mating system common in mammals) and the less common polyandry (one female, many males; found in a

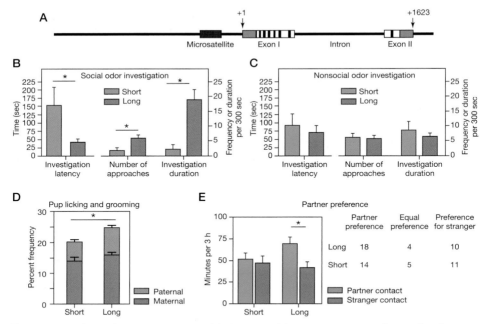

Figure 8.2. Individual variation in parental behavior and bonding in prairie voles is related to V1A receptor microsatellite length. (A) V1A gene. The allele length of this gene in short-allele prairie voles is 727 bp; the allele length in long-allele prairie voles is 746 bp. Long-allele and short-allele voles differ behaviorally. (B) Long-allele voles investigate social odors more quickly and longer, but (C) the two types do not differ in response to nonsocial odors. (D) Paternal pup licking is greater. (E) Although both prairie voles prefer familiar partners to strangers, the preference is greater in the long-allele group.

few species of birds). Both cases differ from promiscuous social systems in that the mating group remains a continually interacting social unit, at least through the breeding season and the rearing of offspring. Both differ from monogamous social systems in that direct care of the offspring is the responsibility of one sex (females in polygyny, males in polyandry), although the other does contribute indirectly by guarding the mate and offspring and, in some cases, the territorial resources for them.

Because polygyny does not evolve from recruiting parental care (or in polyandry, maternal care after egg laying) into the social unit, other factors must be responsible for the evolution of these social mating systems. For polygyny, two factors are generally thought to stimulate its evolution. One is having female choice underlie mating decisions (see Chapter 7). The second is restricted resources, such as the number of high-quality territories, available. If female reproductive success is higher when mating with a male who already has female partners on a high-quality territory than when mating with a single male on a low-quality territory, then polygyny is favored. One factor in weighing those costs is whether active involvement by the male in offspring care significantly increases the reproductive success of both males and females. Intense paternal involvement in offspring care is difficult to support in a polygynous system, because the high aggression needed to defend resource sites and access to females is incompatible with the care of young, and the distribution of male attention across the young of different mothers dilutes the effectiveness of the male's contribution to any one mother. In mammals, this is generally not a concern, as internal gestation and then nursing by the female define that taxon and preclude males from major portions of offspring care. In contrast to this, male birds have ample and important opportunities to aid in offspring care, including brooding eggs and feeding young captured food or crop milk. As a consequence, polygyny is the most common social bonding system in mammals, whereas monogamy is the most common in birds.

The evolution of polyandry is more difficult to understand, especially the so-called classic polyandry seen in several species of birds in which a female mates with several males, each of which has his own nest populated by her eggs. This seems at odds with Bateman's principle, the classic axiom that states that females invest more energy in reproduction than do males and hence are the limiting factor in reproduction (see Chapter 7). Note, however, that intense paternal care, as seen in many birds as well as the pipefish discussed in Chapter 7, counterbalances the male–female asymmetry in gamete investment and leads to social monogamy rather than to promiscuity. One idea is that polyandry is derived from monogamy where paternal care was advantageous. The emergence of polyandry then follows when females become capable of laying more eggs than a single male can successfully care for. Reasons could have little to do with reproduction per se, such as high nest predation selecting for very high egg-laying capacity.

Other instances occur in which multiple males mate with single females because access to females is severely limited. Eusociality (discussed in detail below) is similar to classic polyandry in that a single queen monopolizes matings from multiple male drones as other females in the colony are sterile or reproductively suppressed. A stranger case occurs in deep-sea anglerfish (family Ceratiidae). In 1925, C. Tate Regan reported that what had been thought to be parasites or juveniles attached to female anglerfish were, in fact, dwarfed males that had attached to and fused with the side of the female. Adult males in Ceratiidae species are born with stunted digestive systems that render them unable to eat. They have expanded olfactory systems that are believed to be used to detect female pheromones and specialized jaws that they use to attach to the female. Males secrete an enzyme that digests his and her skin at the attachment to aid fusion between the two individuals. Males gradually atrophy until there is little else but their gonads remaining, which periodically release sperm in response to a female endocrine signal. Depending on the species, females can have one to four dwarf males attached, making this a case of either extreme monogamy or extreme polyandry.

Both polygyny and polyandry are marked by behavioral and morphological features beyond the social bonding characteristics. Most obvious is the evolution of extreme sexual dimorphism in body size and ornamentation. The classic example is seen in the polygynously mating elephant seal. Males are up to three times larger than the females and have large, specialized nasal expansions (the short, thick male "trunks" that give these seals their name) that aid their production of very loud vocal displays. High levels of intraspecific male aggression are also characteristics of polygyny. In polyandry, females take on these traits. Female jacanas, for example, are larger, more conspicuously ornamented, and more aggressive than the males. This carries over to the odd mating system of anglerfish. In the largest species, *Ceratias holboelli*, females can reach >1 m in length, whereas males are ~15 cm long at most.

Social Bonding versus Sexual Fidelity

Cases of pair bonding, social attachment, and biparental care in animal mating systems were called "monogamy" by analogy with human monogamy, which assumes mating with a single individual within the pair bond. The use of molecular genetic methods in the 1990s to explore paternity in nesting birds revolutionized the understanding of such mating systems by showing that, just as in humans, male–female pair bonding may be social monogamy, in the sense that both parents live together and contribute to offspring care, but it certainly does not guarantee sexual fidelity. About 90% of bird species are classified as monogamous. A review by Simon C. Griffith, Ian P.F. Owens, and Katharine A. Thuman noted that <25% of monogamous species have offspring that were found to be 100% derived from the bonded pair. In the remaining species, on average 18% of the broods contained at least one egg fathered by a male other than the female's pair-bonded mate. The rate of infidelity varies greatly among species. At the high end, 86% of reed bunting broods contain eggs not fathered by the paternal care giver, and 55% of all offspring are the result of extra-pair copulations. Extra-pair paternity has also been found in reportedly monogamous primates and canids. Clearly, social monogamy, or pair bonding, is not synonymous with sexual fidelity on the part of either males or females.

The review by Griffith and colleagues suggested that there may be no single explanation for the phenomenon but that, in general, variation in the rate of extra-pair matings across species or families correlated best with two factors. One is the need for paternal care: As the necessity for paternal care decreases relative to maternal care (e.g., for species that breed under conditions of abundant resources), extra-pair matings increase. The second is adult life span. Extra-pair matings are higher when mortality rates are high and adult life span is short. In both cases, the link between social bonding and sexual monogamy is weakened as the risks associated with breaking the bond decrease. Variation on the population level or among closely related species appears to be related more to immediate ecological conditions such as population density. Here, the opportunity to express the tendency guides the behavioral variation.

Far less systematic work has been done on parentage patterns in polygynous and polyandrous species. There, as originally for monogamy, the assumption has been that mating occurs within the social group: Social fidelity equals sexual fidelity. In fact, the evolution of polygyny and its behavioral and morphological characteristics is thought to depend on the fact that only a few males are responsible for nearly all the matings in the population. The revelations that molecular genetic methods brought to monogamous systems represent a cautionary tale for other social bonding systems.

PARENT–OFFSPRING CARE AND BONDING

Successful reproduction depends first on bringing gametes together and second on the resulting zygote becoming old enough to successfully reproduce. Despite this two-stage process, most organisms contribute little or nothing to ensure the successful maturation of their offspring. Once eggs have been fertilized by sperm, the contributing females and males abandon them, a pattern found in many organisms such as fish that release thousands of gametes for external fertilization. Females, or sometimes males, may build nests or other structures for the deposition of fertilized eggs prior to abandoning them. In some cases, largely in insects, the incubation site is provisioned with food for the emerging larvae. In both of these situations, there is no behavioral interaction between parents and offspring. In many mammals and birds, as well as some other animals, in which offspring are unable to fend for themselves, an extended period of behavioral interaction does occur. This parent–offspring bond is the basis of social monogamy, but it also occurs between offspring and the maternal parent in other mating systems. Paternal care in general is much more rare. It is useful to remember that parental bonding and parental care are not exactly the same. Many female mammals and birds show parental care when they are in the appropriate physiological state, but do so indiscriminately toward any juvenile. Bonding is said to occur when the maternal or paternal care is directed only toward the parent's own offspring. In addition to basic parental behavior, it requires the ability to recognize particular individuals as related, direct care only to them, and reject, often aggressively, juveniles of other parents.

Neural and Hormonal Changes and Parental Behavior

As for other vertebrate social behaviors, parental behavior involves an interaction of circulating sex steroid hormones preparing the body and the brain for the task, and brain peptides that serve a dual function—activating body functions related to nurturing and modulating brain systems to change the motivational state and behavioral output of the parents. Comparing examples in birds and mammals shows that, despite some differences in the hormone and neuromodulator patterns, there are fundamental processes apparent. These provide insights into the constraints and opportunities inherent in the interaction of physiology and behavior.

Parental Behavior in Birds

Biparental bird species go through a sequence of behavioral shifts during a breeding season. A general pattern, to which there are exceptions, is as follows: Before mating, males are aggressive toward other males and advertise themselves through singing or visual displays, and females interact with males through their own courtship solicitation behaviors. As bonding and mating begin, the males and/or females build and guard nests. Once the female begins to lay eggs, her behavior shifts into brooding mode, in which a strong tendency to remain in stereotyped positions huddled over the eggs manifests itself. Where males also incubate eggs, their behavior similarly changes. Both may leave the nest and forage, but the active advertisement displays and aggression on the part of the males, other than to defend the nest site, generally abate. Brooding behavior remains until the eggs hatch, at which point males and females are more apt to leave the nest to forage. Either or both parents may feed the chicks, by gathering food and returning it to the nest or by producing "crop milk," a secretion from specialized throat glands that both males and females in several species can produce. Some of the earliest ethological studies demonstrated that behavioral interactions between chicks and parents stimulated the feeding behaviors. The stereotyped gaping behavior in chicks exposed visual stimuli that trigger parental feeding; conversely, signals produced by returning parents, such as visual markings or vibration of the nest as they land, trigger the chicks' gaping response.

The shift in behavioral profile from premating courtship and aggression to post-mating egg incubation and hatchling care coincides with changes in sex steroids and in the peptide hormone prolactin (Fig. 8.3). Experimental studies show how these hormones regulate the two phases. The early phase is characterized by high levels of gonadal steroids—estrogen and progesterone in the females and testosterone in the males. Without these hormones, females will not be receptive or build nests, and males will not display, court females, or act aggressively toward other males. As eggs are laid, gonadal steroids rapidly decline in both sexes, whereas circulating levels of prolactin rise. Prolactin stays elevated throughout egg incubation (brooding) and care of the offspring, but gradually declines. Once prolactin reaches baseline, offspring care generally ceases. For example, in ringdoves the mother's prolactin declines to baseline in ~20 d after the eggs hatch. At that point, she stops feeding the hatchlings. Male doves continue to secrete prolactin for longer than this and continue to feed the chicks. At the time female prolactin falls, she becomes receptive, and the males and females begin courtship and nest building again as gonadal steroids rise. Both gonadal steroids and prolactin are important for physiological regulation commensurate with their behavioral roles. Sex steroid hormones are necessary for seasonal activation of gonads and gamete production; prolactin stimulates seasonal development of crop milk glands (in species like ringdoves that have them) and production of crop milk. Thus, in both phases, hormones are critical for both the physiological activation of body systems necessary for that reproductive phase and for the behaviors that use them.

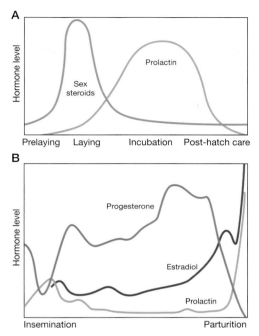

Figure 8.3. Hormone patterns move through similar phases in birds (*A*) and mammals (*B*) to track behavioral states.

As in the case for male–female bonding, there are significant species differences in brooding behavior. Nest parasites such as cowbirds, which lay eggs in the nests of other bird species and abandon the eggs and young to the foster mother's care, show no inclination for incubating eggs or attachment to nests or offspring. The reason is not a lack of prolactin. Cowbirds, like all birds, experience a rapid rise in circulating prolactin upon laying eggs. Presumably, as in the case for male bonding and vasopressin, the answer to these species differences must lie in differences in the expression of brain receptors for prolactin. As yet, such differences have not been explored.

Parental Behavior in Mammals

Paternal behavior is relatively rare in mammals, but with lactation as one of the defining features of mammals, maternal care is universal. Paternal care coincides with male–female bonding and monogamy. Patterns of maternal care and mother–infant bonding vary somewhat among mammals, but in the vast majority it is marked by a radical switch, triggered near birth, in how females react to infant pups (Fig. 8.4). Female rodents that have never given birth at best ignore the presence of pups,

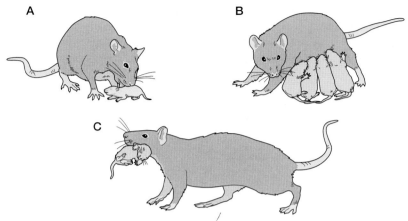

Figure 8.4. Rat maternal behavior consists of (A) licking pups immediately after birth and during the early postnatal period, (B) huddling over pups for nursing, and (C) retrieving pups that stray from the nest. The first behavior is driven by olfactory cues; the second by a variety of olfactory, somatosensory, and thermal cues; and the third by acoustic cues (ultrasonic pup distress calls).

and in many cases find the presence of pups, or even the odor of newborns, aversive. Near the end of pregnancy, however, females become attracted to pups and strive to be near them, which is similar to the "broodiness" that strikes birds as egg laying begins. In experimental studies, foster pups placed in the vicinity of near-term females will trigger pup retrieval behavior, huddling, licking, and other maternal behaviors in the female. Therefore, just as in birds, mammalian parents undergo significant shifts in behavior and motivation, and just as in birds, changes in gonadal steroids and in peptide hormones and neuromodulators mediate the behavioral shifts and the physiological changes necessary for each stage of the reproductive cycle.

Mammalian females experience a dramatic change in both gonadal steroids and peptides around the time of birth, which coincides with the shift to parental care (Fig. 8.3). In rodents and most other mammals, pregnancy is marked by high levels of progesterone and gradually increasing levels of estrogen; at birth, progesterone drops dramatically and estrogen rises rapidly. In Old World primates (including humans), both steroids are high during pregnancy then drop rapidly at birth. Although there are interesting species differences in the effects of steroid hormones on maternal behavior, the most consistent finding across all mammals is that high levels of steroids (especially estrogen) during pregnancy plus low levels of progesterone postpartum trigger maternal behavior. In rodents, having high estrogen after giving birth also stimulates maternal behavior. Paternal behavior, being less common in mammals, is less well studied and its hormonal correlates more variable across species. However, a common finding in rodents is that a male's testosterone level is high during his mate's pregnancy (and coincides with increased male aggression and nest guarding), then drops when the female gives birth and male parental behavior toward the pups (if present in the species) increases.

As gonadal steroid hormones decrease at or near birth, select peptide hormones rise. One important change is a rapid rise in prolactin. In birds, this hormone stimulates secretion from crop milk glands; in mammalian females, prolactin stimulates milk production from mammary glands and as such is an important physiological regulator of the defining characteristic of mammalian maternal care. The role for prolactin in bird parental behavior is clear, but for mammals less so. Facilitating effects have been reported in some studies, but across species and conditions the role of prolactin in females or males is less clear. More important in mammals are the effects of two other peptide hormones and brain neuromodulators with different effects in males and females: vasopressin and oxytocin.

The role of vasopressin in male–female bonding was discussed above. Vasopressin also stimulates male parental behavior in the minority of rodent species that have social monogamy. Vasopressin shows few reliable effects on female behavior, however. Instead, the closely related peptide, oxytocin, has consistently been found to stimulate maternal behavior in mammals. Oxytocin, released by the pituitary gland, increases rapidly in the mother at birth. It stimulates smooth muscle and is important for stimulating uterine contractions at birth, as well as milk release thereafter. In the brain, oxytocin acts as a neuromodulator. Elevated levels there increase the maternal behavior that occurs in late pregnancy and immediately after giving birth. The effect of pregnancy on rat maternal behavior can be mimicked by treatment with oxytocin. Treating virgin rats with oxytocin switches their behavior: They are attracted to pups, tolerating (in fact, seeking) proximity to them. Interestingly, oxytocin release is triggered in females by nipple stimulation produced by infants suckling. This behaviorally induced release is part of the mechanism for the reflexive lactation that follows. However, it may also be part of a continuing infant–mother bonding reinforcement during the prolonged nurturing phase that many mammals have.

How oxytocin (or, for that matter, the other behaviorally important peptides) exerts its effects on maternal behavior is still unclear. Oxytocin may mediate this in part by changing perceptual systems so that infant signals are processed differently. One of the sites of oxytocin (and vasopressin) action in the nervous system is the olfactory bulb, the first station of odor processing in the brain, suggesting that the peptide may change how odors are processed and perceived. This could be important because many of the behavioral interactions between infant and mother that are part of mammalian maternal care enhance the exchange of olfactory signals; and olfactory cues are crucial to the maternal and infant recognition that is the foundation of infant–mother bonding.

A nearly universal maternal behavior, commonly seen from rodents to carnivores to ungulates, is that immediately after birth mothers vigorously lick their newborns, which cleans the infant of the mother's amniotic fluid as well as stimulates its breathing and waste elimination. This provides ample opportunity for the mother to become familiar with the infant's odors. For mothers in species such as sheep, where there is strong mother–infant bonding, exposure of the mother to the offspring's odor during this critical period immediately after birth is essential for the mother to form a specific attachment to her own young. Conversely, infant mammals use odors to recognize their mother. Rat pups appear to imprint on the mother's odor while still in utero. If a distinctive odor, such as lemon oil, is provided in the amniotic fluid, pups are attracted to it after they are born, and, in fact, the odor preference persists long into the juvenile period.

Somatosensory signals, from tactile stimulation associated with mothers licking pups and pups contacting the mother's ventrum to thermal cues that change as pups begin to thermoregulate and increase their body temperature, are also important to maintain the mother–infant bond and regulate the expression of maternal behavior. Nevertheless, recognition and bonding to particular individuals or litters, as well as the approach and preference for pups that female rodents show only after giving birth, are strongly dependent on odors. In addition to any other brain effect it has, it may be that oxytocin changes a mother's olfactory processing in a way that her perceptual world is significantly altered at birth: The salience or emotional connotations of pup odors suddenly change. Whether this is, in fact, the case awaits experimental studies. Oxytocin also interacts with the brain's reward system. Oxytocin may thus orchestrate a suite of coincident brain changes so that social signals become more detectable and salient, while at the same time these signals, meaning the presence of pups, become rewarding rather than aversive.

Auditory cues are also important parts of mother–infant behaviors. Rats and most other rodents communicate acoustically in the ultrasonic range (i.e., at frequencies higher than humans can hear). Distress calls from pups are ignored by virgin females, but after giving birth, these calls are powerful stimulators of female attention and trigger searching and retrieval behavior. Robert C. Liu used sophisticated electrophysiological and analytic methods to explore the sensory correlates of this behavioral change. Sounds of all kinds are coded by the auditory cortex in mammals, a temporal lobe area that is the first cortical recipient of auditory information ascending through the lower auditory pathways. Liu compared the cortical response to pup distress calls and to nonbiological control sounds in naïve virgin female mice and mice that had given birth. In mothers, but not in naïve females, neurophysiological responses in the auditory cortex were able to reliably follow sequences of pup calls at increasingly faster repetition rates up to the naturally occurring pup call repetition rate of five calls per second. Furthermore, the latency to the point at which sufficient neural activity occurred to detect a call (as determined by signal analysis programs) was significantly shorter in mothers, as was the magnitude of the activity when that peak was reached (Fig. 8.5). Further analysis showed that the information in the mother's neural coding provided more reliable discrimination among call signals and provided it at an earlier point than in naïve females, and that more areas within the auditory cortex contained such high-detection, high-discrimination coding. Differences between mothers and

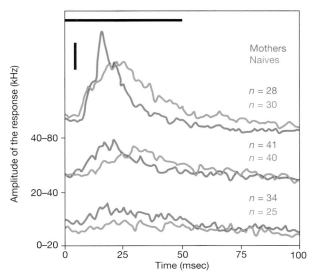

Figure 8.5. Neural responses in the auditory cortex to ultrasonic signals differs in mother and naïve female rats. Amplitude of the response over time (x-axis) to a pup distress call (64-kHz median frequency; 50-msec duration; 65-dB sound pressure level, dB SPL) is plotted as an averaged, smoothed function. The vertical black bar equals a response magnitude of 50 spikes per second, and the black horizontal bar marks the signal presentation. The responses are shown at three different sites in the primary auditory cortex coding different characteristic frequencies (shown on the y-axis). The difference is most pronounced in the cortical area tuned to the frequencies characterizing pup distress calls.

virgins were only apparent for pup distress calls; the processing of control acoustic signals did not differ. What Liu's work shows is that, in fact, a female's perceptual world *does* change after giving birth to pups. We cannot know how mice really perceive the sounds around them, but the results do suggest that infant calls now become more salient and more sharply discriminated than other sounds in the female's environment, changing the auditory landscape in which the female mouse lives. As yet, the physiological factors behind this change are unknown.

The Trade-Off in Courtship versus Parental Care

The change from mating to parenting is not a necessary dictate of physiological cycles but a response to the challenges of balancing the expression of each to maximize reproductive success in individual males and females. The functional reason behind the shifting balance of courting and parenting is that the behavioral and physiological suites of characters defining these two phases of reproductive social behavior are in conflict. Consider the simple problem of time budgeting. Time spent searching for a mate, fending off suitors and rivals, and fighting to establish a territory to attract a mate is time away from brooding eggs, feeding young, and defending nest sites from predators. Furthermore, intraspecific aggression and expelling rival conspecifics from resources is incompatible with tolerating close proximity to other individuals and sharing those resources with them.

The cost–benefit considerations are quite different in birds and mammals. In birds, males can make a substantial contribution to the successful rearing of offspring given the external development of eggs and the possibility for sharing in the feeding of young. Hence monogamous pair bonding and biparental care are common, and a shift from courting to parenting is seen in both sexes. In mammals, internal fetal development and lactation put the onus on females and vastly reduce the need, and opportunities, for males to help after mating. What then becomes important for male reproductive success is mating with as many females as possible, as in promiscuous species; in polygynous species, this instead dictates constant defense of the female harem from other encroaching males. A behavioral and physiological switch marks females, but in most cases, males remain aggressive or in courtship mode without switching to a less aggressive parenting phenotype. In both birds and mammals, testosterone is the key hormone setting the switch between the two behavioral states.

It is important to remember that pair bonding and parental care do not block mating in either males or females. Extra-pair copulations discussed above happen while bonded pairs have already established a parental relationship. Superb fairy-wren (*Malurus cyaneus*) males engage in courtship and parental care simultaneously, with a very high level of extra-pair copulations. Testosterone treatment changes the balance, increasing courtship behavior at the expense of parental behavior, with the biggest change an increase in courtship behavior toward the bonded female. In some Old World primates—humans being the prime example—mating behavior continues during the exceptionally long period of biparental offspring care. Understanding species differences in mating systems and bonding is best done from a functional perspective, informed by mechanistic considerations. Given life history characteristics, what represents the best overall balance for reproductive success, and how does this differ in males and females? Given that balance, how are the conflicting hormonal and physiological factors regulating courting/aggression and parenting/tolerance reconciled to yield the particular strategy used by the species?

Behavioral Experience Can Override Physiology

The wealth of data on the physiological mechanisms of reproductive social behavior masks the fact that behavior itself is a powerful driver of social behavior, so much so that experience can substitute for physiological mechanisms. In nature, of course, hormonal state, reproductive physiology, and mating behaviors normally co-occur. In the laboratory, these can be dissociated with interesting results. Virgin females can be induced to express attraction toward pup odors and express maternal behavior such as nest building and pup retrieval by gradually exposing them to pup odors in repeated trials over several days. On the mating side, castration severely decreases, if not eliminates, male courtship and mounting. In rodents, the strength of this experimental effect depends on the mating history of the male. Castrating male rats that have never been exposed to females completely eliminates their reproductive behavior. They do not court or mount females after the testosterone-eliminating surgery. But blocking testosterone in male rats after they have had even one copulatory experience with a female has little immediate effect. Even without testosterone, such males will continue to be attracted to females, produce courtship behavior, and successfully mount and show intromission behavior for days or weeks. In both cases, previous behavioral interactions somehow substitute for the hormonal control of new reproductive behaviors, or reconfigure the brain so that it can operate in the absence of those hormones.

In Chapter 7 we noted Daniel Lehrman's classic work showing that behavioral interactions induce changes in reproductive physiology and hormone levels, which, in turn, trigger additional behaviors. This is another example showing that behavior has as much of an effect on endocrinology as endocrinology has on behavior.

ALTRUISM AND COOPERATION

Reproduction and parental care are examples of cooperative behavior. Males and females cooperate in the behaviors leading to gamete exchange; biparental species work together to care for offspring. Parental behavior is also altruistic in the sense that mothers and sometimes fathers work to ensure that their offspring have protection, nests, and food by providing resources that they themselves could otherwise use. These behaviors are understandable from a fitness perspective: Cooperation and providing resources directly increases an individual's fitness by ensuring gene transfer to the next generation (the very definition of fitness), so in a sense this behavior is not really altruistic at all. What puzzled evolutionary biologists was how to deal theoretically with cooperative or altruistic behavior that appeared beyond the immediate parental–offspring bonds. Cooperative behaviors are as diverse as pack hunting, producing alarm calls, and cooperative breeding behavior that can range from group nesting to reproductive social systems in which nonbreeding adults help breeding pairs care for offspring. The latter case looks altruistic: Individuals are deferring their own reproductive effort to help others, and in the strictest interpretation are decreasing their own immediate individual fitness by increasing the fitness of others. How can such a behavior evolve?

The solution to the evolution of cooperative behavior lies in W.D. Hamilton's classic articulation of the concept of inclusive fitness. Behind the concept of inclusive fitness is the idea that what matters for evolutionary processes is the increase in overall gene frequency, not just the survival of an individual. Inclusive fitness therefore reflects an individual's fitness from its own offspring plus the number of offspring equivalents it can add to the population by supporting other individuals. Genetic traits will thus be favored if an individual's actions favor its own survival, its offspring's survival, or that of relatives sharing its genes. Hamilton's ideas and the concept of "kin selection" were reviewed in Chapter 2 along with its implications for altruistic behavior. We will also return to it later in this chapter.

The theory of inclusive fitness predicts that altruistic cooperation is more likely to emerge when the performer and recipient are closely related, a prediction that has repeatedly been confirmed. Altruism and cooperation will only evolve in unrelated individuals when there is a direct benefit to the individual performing the altruistic act (e.g., when reciprocity is expected and enforced) or when some other net gain is acquired, on average, by cooperative individuals above what each could acquire individually. Although bringing into question how altruistic altruism actually is, inclusive fitness considerations elegantly bring cooperative social behavior back into the consideration of individual fitness that is the foundation of all modern evolutionary biology. Cooperative hunting can evolve if the result is a net increase in food availability per individual in the pack. Alarm calling can evolve if it either reduces the predation risk of the caller or enhances the survival of related individuals at little increased cost to the caller. Cooperative breeding is more complicated. It can evolve when the helpers enhance the survival of the genes they share with the targets of their help, but only when there is some extrinsic reason that limits the helper's ability to reproduce itself, as ensuring the survival of one's own offspring has a higher fitness benefit than does ensuring the survival of one's siblings, nephews, or more distantly related relatives. Cooperative breeding should therefore be relatively rare and occur under conditions that preclude all individuals from breeding effectively.

Cooperative breeding has been most thoroughly studied in birds, where it follows the predictions based on inclusive fitness considerations. It is seen in ~3%–4% of avian species. Most commonly, helpers are nonbreeding relatives of the breeding pairs they help, often older offspring (the closest relatives to the young benefiting from the additional help). The most common ecological factors in cooperative breeders across avian taxa all relate to a shortage of breeding opportunities, either because of a lack of available territories, breeding sites, or mates. There is strong evidence that cooperative breeding really represents a holding pattern. In the absence of an opportunity to breed, helping relatives care for offspring represents a marginal increase in fitness over doing nothing. It is universally the case, however, that helpers will leave and adopt independent breeding status if the opportunity arises.

Life history considerations combine with ecological conditions to predict the evolution of cooperative breeding. Cooperatively breeding bird species have low annual mortality rates and generally low clutch sizes. They tend to occupy stable, relatively resource-rich habitats, where they are sedentary rather than migratory; cooperative behavior is three times more likely in tropical birds than in higher-latitude species, for example. These factors make it more likely that a helper will eventually have the opportunity to successfully breed on its own, as well as exploit any direct benefit it gained through its experience as a helper. In fact, in some cases, the main benefit to helpers is that they increase their chances of gaining a territory in the future.

Cooperation among Siblings

Kin selection predicts that altruistic behavior should be biased toward close relatives and that the closer the relatedness the more likely altruism will be favored by selection. Specifically, as we reviewed in Chapter 2, Hamilton's rule even predicts when altruistic behavior will be favored by selection, when the cost, in terms of units of fitness, to the actor (c) is less than the benefit accrued by the recipient (b) discounted by the degree of relatedness (r): $c < r * b$ (see Fig. 2.3). Rearranging the equation, we see that the net benefit to the altruist is $rb - c$.

Because of its dependence on r, altruistic behavior should be more likely among siblings than among other classes of individuals. In some cases, cooperation among siblings functions to enhance mate attraction. Birds, for example, form coalitions to court females. In several species of manakins, two or more males synchronize their songs as well as their dances in an attempt to attract females, but only the alpha male mates (Fig. 8.6). It had been thought that these males were brothers, but molecular genetic analysis rejects that hypothesis. As with cooperative care, the benefit obtained by the beta males is the higher probability of someday becoming an alpha male. A recent study by Alan Krakauer, however, shows that in turkeys the band of brothers involved in courtship coalitions does gain a kin selection advantage. As is the case with manakins, only the dominant male mates. Microsatellite analysis shows that the average relatedness among the cooperatively courting males is 0.42 (r), close to the 0.5 value expected of full siblings. Although solo-courting males sometimes mate and fathered an average of 0.9 offspring, dominant males in a courting coalition had an average of 6.1 more offspring (b) than a solo male. Subordinate males do not mate; thus they sacrifice the average of 0.9 offspring (c) they would garner if they were solo-courting males. According to Hamilton's rule, $rb - c = 1.7$, which is a net benefit to subordinate males who forsake mating to enhance the mating success of their close relatives.

In some cases, altruistic sibling interactions can occur early in development. Toad tadpoles, for example, associate with siblings in forming schools. Toad tadpoles are distasteful, but eating one rarely results in the death of a predator. But if ingestion of a tadpole teaches a predator that individuals from this school are distasteful, the predator will avoid the other tadpoles in the school. Thus, the death of a single tadpole can benefit the fitness of a large number of close relatives. Ronald Fisher suggested such a scenario to explain the evolution of distastefulness as an antipredator mechanism.

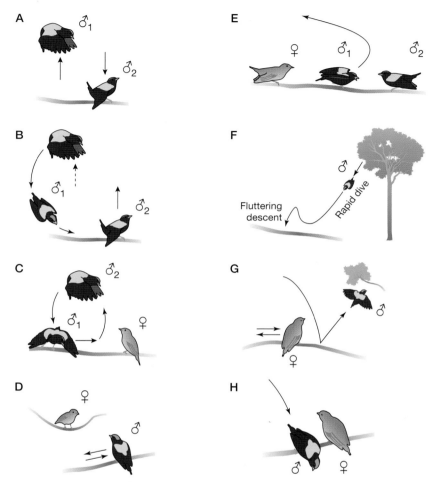

Figure 8.6. Cooperative sexual displays by male lance-tailed manakins. (A) In a typical cooperative courtship sequence, two males—an alpha male (male 1) and a beta male (male 2)—land within a few centimeters of each other on a display perch and alternately leap into the air, hover, and then land at the same location. During this up-and-down display, the males produce a dance-call vocalization. Females are usually in the area but not on the display perch during the display, but the dance attracts a female to the display perch, at which time the display ends. (B) Sometime males add a swooped descent to the dance in what is known as the quick-down variant of the up-and-down display. (C) In the leapfrog display, with the female on the display perch, male 2 leaps into the air, male 1 moves into his place on the perch, and male 2 lands near where male 1 had been perched. The males continue this dance, with each male facing the female as he hovers above her. (D) The back-and-forth display is performed by a single male in front of a female in nearby vegetation. (E) Preceding an eek display, the leapfrog display leads to a frenzied pace with shorter and faster hops by the males. One male positions himself between the female and his partner male. He then produces an eek call and flies from the perch, eventually followed by the other male. (See facing page for legend.)

The Neural Basis of Cooperation, Tolerance, and Trust

Cooperative breeding behavior is conceptually an extension of maternal care. Just as mother–infant bonding results in the parent sacrificing its resources for the benefit of its offspring in exchange for an inclusive fitness benefit, helpers suppress their own immediate selfish tendencies for the promise of a longer-term gain. Cooperative behavior more broadly, whether driven by kin selection considerations or immediate benefits, has similar traits. In all cases, an individual opting for cooperative behavior must express some of the same traits that are crucial for bonding with members of the immediate family unit. These include a suppression of aggression in favor of affiliation; tolerance for the presence of other, potentially competing, individuals; and reciprocity of the efforts and resources when the circumstances are reversed. Interestingly, cooperative behavior and its related psychological traits may also be linked mechanistically to male–female and to parent–offspring bonding. Recent work suggests that the nonapeptides that play a critical role in familial interactions—vasopressin (or its non-mammalian homolog vasotocin) and oxytocin (or its non-mammalian homologs mesotocin and isotocin)—may, in fact, generally influence cooperative behavior, social tolerance, and trust.

Oxytocin, an important modulator of maternal behavior, in particular appears to have an important influence in both males and females. One feature of familial behavior is the tolerance for social proximity. This kind of behavior occurs in other contexts, such as flocking, shoaling, and herding. Species vary in their tolerance of such group behaviors. Birds in the family Estrildidae all form long-term monogamous pair bonds for breeding and exhibit biparental care, but vary greatly in other aspects of their social organization. Zebra finches (*Taeniopygia guttata*), for example, are highly gregarious, spending their time in large aggregations. James Goodson tested zebra finches of both sexes for their preference to associate with familiar versus unfamiliar birds, and with large versus small groups (Fig. 8.7). Treatment with an oxytocin-receptor antagonist reduced their preference for familiar birds, and their preference for being near large rather than small groups of conspecifics. Conversely, treatment with mesotocin (the avian homolog of oxytocin) delivered via implanted cannulas directly into the brain had the opposite effect: Increased mesotocin increased time spent with familiar birds and with larger groups. The effects were most pronounced in females.

Figure 8.6. (*Continued*) The two males either return to engage in more leapfrog displays or the male that produced the eek call returns to court the female by himself. (*F*) In the swoop display, a male flies to a site 10–20 m away, produces a call, and then rapidly flies back to his display perch, producing a swooshing noise as he flies. (*G*) A male gives a bounce display by flying from a nearby branch to his display perch, where he briefly alights and returns to the branch, appearing as if he is bouncing off the display perch. This display usually precedes a copulation attempt. (*H*) The bow display only occurs after a bounce display. The male faces away from the female, bows his head, and then attempts copulation.

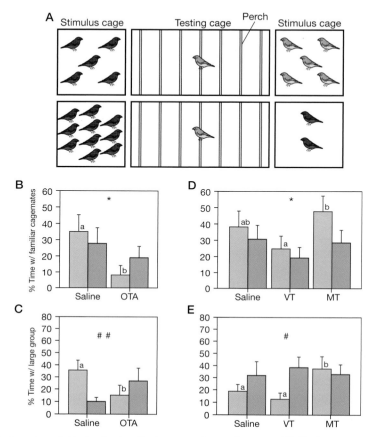

Figure 8.7. Experimental tests of zebra finch aggregation preferences. (A) Choice apparatus in which females or males are placed in the center chamber and their proximity to either of the test chambers is measured. Birds are tested for their preference to be near familiar versus unfamiliar birds in equal numbers, and to be near large versus smaller numbers of unfamiliar conspecifics. Treatment with oxytocin receptor agonist (OTA) reduced time spent with familiar birds (B) and with larger groups (C), with effects most pronounced in (blue) females (males, red). Conversely, treatment with (D) mesotocin (MT), but not (E) vasotocin (VT), increased the preference for familiar birds and larger groups, again in females (D,E).

Goodson followed these experimental studies by examining oxytocin receptor distribution in different Estrildid species. The locations of oxytocin binding sites were the same across all species (and, in general, are highly conserved across vertebrates), but the density of binding sites in one particular brain area, the lateral septal nucleus (a limbic system brain area implicated in social behavior and aggression), differed. Gregarious species had significantly higher binding capacity there than did congeneric territorial species that do not tolerate being in large social aggregations (Fig. 8.8).

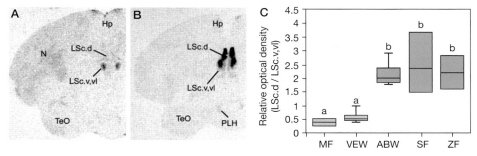

Figure 8.8. Oxytocin receptor binding differs between gregarious aggressively territorial violet-eared waxbill (A) and Angolan blue waxbill (B). Binding in the dorsal lateral septum (expressed here as ratio of dorsal to ventral) is higher in three gregarious species compared to two territorial species of Estrildid finches (C). Abbreviations in photos: LSc.d, c.v,vl indicate divisions of the septal nuclei; Hp = hippocampus; N = nidopallium; TeO = optic tectum; PLH = posterolateral hypothalamus. Abbreviations under bar graphs: MF = melba finch, VEW = violet-eared waxbill, ABW = Angolan blue waxbill, SF = spice finch, ZF = zebra finch.

Some of the most interesting recent work on the role of nonapeptides on social cognition involves experimental studies in a taxon often not considered as a model organism for animal behavior: humans. Working with people has the advantage in that cognitive and social psychologists have devised complicated cooperative behavior experiments probing trust, altruism, and other cognitive processes that are difficult to examine in nonhuman species. The results of these studies (so far done mainly in men) show that oxytocin manipulates social perception and cognition in ways predicted by animal behavior studies. Treatment with oxytocin (using a nasal spray) increases subjects' attention to human faces and increases their gaze toward the eyes. It also enhances subjects' ability to categorize human faces based on the emotional state those faces convey. In addition to changing the perception of social stimuli, oxytocin changes human social decisions. Social cognition researchers have studied economic decisions in a "trust game." Here an "investor" chooses to give an unfamiliar "trustee" an amount of money. The "invested" money is then increased by the experimenter, and the "trustee" chooses to return a larger or smaller amount of money to the "investor," keeping some as a "commission." The game progresses with successive decisions by the "investor" to entrust funds to the "trustee." The game was designed to probe how individuals weight risk and reward, and how interacting individuals balance cooperation–competition in the face of uncertain future rewards. Of interest here is that intranasal application of oxytocin in the "investor" increases the amount of money he is willing to offer the "trustee." Something— usually interpreted as level of trust—is being manipulated by the elevated oxytocin. Human brain imaging studies such as a functional magnetic resonance imaging (fMRI) experiment by Peter Kirsch and colleagues suggest that oxytocin treatment is accomplishing this, at least in part by reducing activity in the limbic system, specifically the amygdala, which is particularly important in assessing negative social and emotional conditions. It is significant that both the perceptual and the cognitive effects are restricted to social contexts. There is little evidence that oxytocin changes perceptual processing of arbitrary visual stimuli, and "investors" do not increase their "investment" when they know the return is determined by a random event like a lottery.

The mechanistic commonality between parental behavior, cooperative social behavior, and prosocial cognitive processes such as trust raises fascinating evolutionary questions. Do complex cooperative social structures evolve out of familial pair bonding and parental care? Or is it the case that parental care and male–female pair bonding are evolutionary specializations of some general mechanism supporting social aggregations? Do changes in social perception necessarily coincide with changes in social cognition, or can they be dissociated mechanistically, functionally, and evolutionarily? Does individual or taxonomic variation in parental behavior predict variation in behavior in other social interactions? Questions such as these show how evolutionary, functional, and mechanistic questions are interrelated.

FROM SELF TO KIN TO SPECIES RECOGNITION

For kin selection to operate, altruistic behaviors must be biased toward close relatives. This does not necessarily entail active kin recognition. Individuals in close proximity are more likely to be close relatives in animals that have low dispersal abilities. Thus, any altruistic behaviors are more likely to be biased toward close kin; this is sometimes referred to as indirect kin recognition. In other animals, however, kin recognition is an active process.

Kin recognition needs to be based on cues that are reliable indicators of relatedness. In some systems, these cues are instantiated by the local environment. Hive odors in social insects, for example, are used to recognize colony mates. In other systems, recognition is based on highly variable aspects of the phenotype or genotype.

In the previous chapter, we discussed the role of MHC genes in mate choice. Owing to their immune function in identifying cells as self or non-self, these genes are under selection to be hypervariable. This also makes these genes ideal indicators of relatedness and predicts that phenotypic cues tied to these genes, such as odors produced in the urine, could be potent kin recognition cues. Recently, Bruce Waldman showed that tadpoles can recognize kin based on MHC variation. The challenge in demonstrating such a phenomenon is that MHC variation among animals of different relatedness will be confounded by other correlated variables. Using tadpoles of the species *Xenopus laevis*, Waldman tested social affiliations in lines of full sibs with different MHC haplotypes. A tip of the tadpole's tail was removed for genotyping, and individuals were then tested for time of associations with tadpoles of the same versus different MHC genes. The results showed that among full siblings, those with the same MHC haplotypes preferentially associated with one another.

Recognizing the phenotypic product of genes is an important mechanism for kin recognition as genetic variation is key to genetic relatedness. The more genes involved in recognition, the more likely they will give an accurate assessment of relatedness. With the exception of the highly variable MHC genes, basing recognition on a single gene might not give similar results. In fact, in theory, a single gene could hijack altruistic behavior for its own advantage.

According to the genic view of evolution proposed by W.D. Hamilton and popularized by Richard Dawkins in *The Selfish Gene*, genes are under strong selection to replicate themselves. Hamilton predicted the existence of supergenes with three properties: (1) a distinctive phenotypic trait, (2) the faculty to recognize the trait in others, and (3) the propensity to direct benefits toward bearers of that trait, even though this entails a fitness cost. Dawkins referred to these imagined genes as "green beard genes," to denote how unlikely their existence might be.

But how unlikely are they? In discussing mother–offspring conflict during human placentation, David Haig pointed out that maternal cells will limit the degree of placentation, whereas genes of the fetus will be under selection to extend it: The fetus wants more resources than the mother wants to provide. Haig suggested that this asymmetry created an arena conducible to a green beard effect; green beard genes in the maternal cells would allow fetal cells with the same green beard genes greater access to maternal resources through more extensive placentation. Not only a theoretical possibility, Haig suggested that such green beard genes exist under the guise of homophilic cell adhesion molecules. Such molecules contain all the prerequisites for green beard genes. Their extracellular domains recognize copies of themselves on other cells, and their cytoplasmic domains are able to initiate actions within the cell. Thus, if the same allelic variant were expressed by molecules on each side of the maternal–fetal interface, there would be the foundation for a green beard effect if this resulted in actions that favored the fetus.

Haig's argument brought the idea of green beard genes from an intriguing theoretical musing to a real possibility. Studies of the social amoeba *Dictyostelium discoideum* by David Queller, Joan Strassmann, and their colleagues demonstrated that green beards were far less fanciful than Hamilton and Dawkins had thought. These amoebae forage singly until food runs out, and then they come together in the thousands and form long stalks that support reproduction in other cells that differentiate into a cluster of spores (Fig. 8.9). As many as 20% of the amoebae die in the formation of these stalks, and cells from different clones readily form chimeric fruiting bodies for reproduction. In this system the stage seems to be set for kin- or genetic-biased altruistic behavior as there is both self-sacrifice and variation in relatedness. These conditions, however, have been hijacked by a single green beard gene. The *csA* gene encodes a cell adhesion protein that has a globular domain, gp80, which binds to the same domain in gp80 proteins of other individuals (Fig. 8.9). If an individual does not have the gp80 protein, as knockout experiments show, it is left behind in the aggregate streaming that leads to formation of the slime mold. Spores from the

A

B

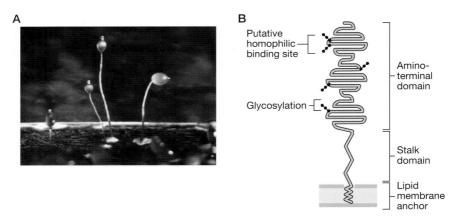

Putative
homophilic
binding site

Glycosylation

Amino-
terminal
domain

Stalk
domain

Lipid
membrane
anchor

Figure 8.9. A green beard gene in an amoeba. (A) Fruiting bodies of the social amoeba *Dictyostelium discoideum*. (B) The *csA* gene (contact site A) codes for *gp80*, a homophilic cell adhesion protein.

resulting chimeric fruiting bodies were 82% wild type because their homophilic binding allowed them to adhere in aggregation streams and to pull each other into aggregates, whereas most knockouts were left behind. The *csA* gene ensures its survival by binding to individual cells that have with proteins from the same gene. Some of these individuals die in the stalk, but others go on to reproduce with a much higher probability than that of the amoebae without the gene. This is not kin selection because the aid is not given on the basis of relatedness but on the basis of having only a single gene in common. Thus, the *csA* gene embraces the criteria that Hamilton developed for a green beard gene, a gene with a variable phenotype that recognizes other identical genes and that results in altruistic actions that enhance the replication of like genes, even in the face of self-sacrifice.

It might be that single green beard genes are more likely to be found in microorganisms than in animals. As suggested by Queller and his colleagues, in multicellular organisms the three functions of a green beard gene would all have to be coordinated through different cells, different tissues, and different organ systems. This complexity might guard the individual and its parliament of genes from being hijacked by a single selfish gene.

So far, we have seen that expression of altruistic behavior can be influenced by a green beard gene, a hypervariable gene complex that is more indicative of overall relationship, such as MHC, or environmentally acquired cues, such as colony odors, which indicate spatially close and thus usually closely related individuals. Some of these mechanisms clearly have a strong genetic basis, whereas others can be acquired through imprinting or other types of associative learning. Underlying all these mechanisms is self-similarity or phenotype matching, a comparison of some aspect of another's phenotype to your own. Dawkins referred to kin recognition based on familiarity as the "armpit effect."

Humans have diverse phenotypes; no two bodies are exactly the same. Our fingerprints, for example, allow near-error-free identification of individuals. Although we can rely on numerous cues to identify others, we seem especially adept at using faces to tag individuality. A recent study mimicking the "tragedy of the commons" illustrates the role that faces have in directing altruistic behavior. Subjects participated in a "public goods game" in which they contributed varying amounts of money to the public good, after which profits were evenly shared. The faces of the other members of the public were shown on a computer screen during the game. Unbeknownst to the subject, however, these were not actual participants at other universities as they had been told, but were sham photos, some of which were morphed in a 60:40 ratio of the subject's own face. The participants were shown photos of unfamiliar faces and the morphed photos. The results showed that subjects were more likely to contribute to the public good as the number of sham participants with morphed faces increased. Altruistic behavior in humans seems to be modulated in some extent by the overall similarity of the face of the potential recipient to that of the subject. Humans, as do a number of other species, sometimes rely on phenotype matching to direct altruistic behaviors.

We have been discussing recognition at the level of kin and genetic relatedness. But recognition also occurs at higher levels of organization. In Chapter 7 we discussed how mate choice can be exhibited between species, populations, and individuals within a population. Selection favoring mate choice at one level can have incidental effects at other levels. For example, selection to favor conspecific over heterospecific mates can result in neural recognition mechanisms that guide females to conspecific mating signals. These recognition mechanisms are templates against which incoming signals are compared.

As we reviewed in Chapter 7, mating signals and preferences for them can diverge in allopatry to enhance species identification. Most pairs of species, however, are not in contact with one another. Yet when challenged with the mating signal of a conspecific versus that of an allopatric heterospecific one, females rarely make a mistake, despite never having been exposed to that heterospecific. The reason might be that animals compare mating signals of all individuals to a template of self, either a learned one or a hardwired one. Support from this comes from studies of the túngara frog.

Michael Ryan and his colleagues constructed five "acoustic transects" that consisted of a series of synthetic calls that varied in similarity from the conspecific mating call of the túngara frog to the mating call of one of five heterospecifics, all of them allopatric with túngara frogs (Fig. 8.10). There were 13 calls in each transect. Females were tested with each call in a recognition test; if a female approached the call, this showed that she recognized it, in error, as a potentially viable mate. The results, presented in Figure 8.10, showed that the strength of recognition decreased predictably with the acoustic distance of the test call from that of the conspecific call. Thus, females generalize in mate recognition; they determine the attraction of a novel signal

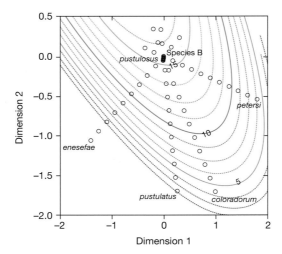

Figure 8.10. Generalization of mating calls in túngara frogs. Acoustic transects between the calls of the túngara frog and five heterospecific frogs are shown in multidimensional space. The contours represent the strength of female recognition of each stimulus estimated by a quadratic smoothing function. Túngara frogs used in these experiments are allopatric with all of the heterospecifics tested. Nevertheless, heterospecific calls are not recognized as signifying appropriate mates, and the strength of recognition scales smoothly from the conspecific call through intermediate calls to the heterospecific calls, suggesting that females generalize from a template that recognizes the túngara frog call.

by comparing it to some representation of the conspecific signal (remember, females do not produce a mating call), and thus exhibit a type of phenotype matching. Recognition from the level of cells through individuals to species might sometimes follow some very general principles of self-recognition.

EUSOCIALITY

On one extreme of female mating patterns is monogamy. Females bond to one male, and most females in a population reproduce during their lifetimes. On the other extreme is eusociality. In a large colony, one female (or a very small number) is responsible for all reproduction, mating with a small number of males (sometimes one). Most individuals never reproduce, and, in fact, large numbers of offspring may be sterile. Eusociality also represents an extreme case of altruistic, cooperative reproduction: The great majority of individuals sacrifice their own reproduction to support the reproduction of others. Eusociality represents an extreme case of social inequality, with many individuals serving the (reproductive) needs of a very few. Eusociality is widespread among the hymenoptera (ants, bees, and wasps) and the isopteran termites. The only eusocial vertebrates are two species of mole rats.

Eusociality is characterized by a complex, hierarchical behavioral organization, marked by cooperative reproduction, within a group of genetically related individuals. The great majority of individuals within the social group defer their own reproduction to support the reproductive effort of a reproducing female, or queen. There is a division of labor within the colony with specialized castes, some of which are sterile. Generations overlap, and caste duties can either be fixed or individuals can move from one function to another as they age. Caste members differ significantly morphologically and physiologically, as well as behaviorally (Fig. 8.11). Honeybees provide a well-studied example of eusocial insects. A small number of males, or drones, mate with the queen. Honeybee queens store sperm for later fertilization of enormous numbers of eggs. Other female members of the colony serve as workers of different kind: foragers, soldiers, and nursery workers. Such workers in many eusocial insects move from in-hive duties, such as tending eggs, to out-of-hive duties, such as foraging, as they age. Drones, the males, have large eyes (presumably to aid in the mating flight behaviors), but they have an ovipositor rather than a stinger. Workers are smaller overall, but are more aggressive and have stingers. Foragers, which must fly long distances to find food and return to the hive, have enlarged mushroom bodies, a brain area in insects that is implicated in sensory integration and memory formation.

As with other cases of cooperation and altruism, the explanation of the evolution of this behavior lies in a consideration of inclusive fitness. Genetic relatedness among individuals within the group is essential for eusociality to be sustained, just as it is a common element in less extreme examples of cooperative breeding. In the case of eusocial hymenopteran insects, their peculiar system of sex determination is thought to be an additional factor in the emergence of eusociality.

Genetic sex determination was discussed in Chapter 6, where the haplodiploid determination system was mentioned. In this system, sex is determined by the number of chromosomes: Diploid individuals are female, haploid individuals are males.

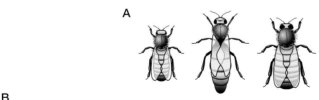

A

B

Sex	Daughter	Son	Mother	Father	Full sister	Full brother
Female	1/2	1/2	1/2	1/2	3/4	1/4
Male	1	N/A	1	N/A	1/2	1/2

Figure 8.11. (A) Worker, queen, and drone honeybees differ in size and other morphological features in an example of physical differences among eusocial castes. (B) Genetic relatedness among relatives differs depending on the method of sex determination. This table shows shared gene proportions in haplodiploid sex-detemination system relationships.

Both males and females are fertile, and during mating, males contribute all of their genes, whereas females contribute half of theirs. Females result from fertilized eggs, males from unfertilized eggs. This presents some odd genetic relationships. Males have no fathers and cannot produce sons, but can produce daughters whose sons are their grandsons. Females can produce both sons and daughters. Most importantly, the degree of genetic relatedness among females is different from that in the more usual genetic sex determination (Fig. 8.11). Females are more related to their sisters than to their mother or their own offspring. This is because each sister inherits all of her father's genes and half of her mother's. If parentage is from the same drone and queen, sisters share three-quarters of their genes. On average, they share only one-half of their genes with their mother, and pass on only one-half of their genes to their own daughters. By this quirk of genetics, a female thus gains more fitness by raising a sister than by raising a daughter or a son! This oddity is thought to predispose haplodiploid species toward cooperative reproduction, whereby female workers forgo producing sons or daughters and instead work to raise younger generations of siblings.

In fact, all eusocial hymenoptera have this mode of sex determination. Eusociality has evolved independently in other insects without haplodiploid sex determination, but which share with the eusocial hymenoptera extreme degrees of genetic related-ness within a group. Species of aphids (order Hemiptera) and thrips (order Thysanop-tera), for example, express eusociality. In those species, some reproduction is asexual, giving rise to sterile soldiers that are clones of the reproducing queen, thereby sharing 100% of her genes and the genes of their sisters. What remains puzzling is how euso-ciality could have arisen in the only two vertebrate species that show it, the naked and Damaraland mole rats, which determine sex via the standard mammalian X and Y sex chromosomes.

Sex determination methods predispose the evolution of eusociality, but only to the extent that they give rise to an asymmetry in relatedness. One other consideration is that competition among individuals within a group must be more costly to individuals than is cooperation. A similar statement could be invoked for the evolution of cooper-ative pack hunting or herd predator defense. Eusocial species often gather and store resources and defend these resource and reproduction sites against intruders: mounds and hives in hymenoptera, galls in aphids and thrips, burrow systems in naked mole rats. If individuals are incapable, or less likely, to survive on their own because, for example, they cannot defend such sites as solitary individuals, on balance it is better to cooperate with related individuals than to compete with them. Even if, as often hap-pens, nonreproductive soldiers sacrifice themselves in the defense of the group home, inclusive fitness considerations make the sacrifice of a minority a sustainable strategy if a sufficient majority survives to maintain the genetic endowment.

A second factor often not discussed is that some level of coercion and manipu-lation by individuals at the top of the hierarchy is needed to sustain the inequities inherent in eusocial groups. This can come at multiple levels. Consider, for example, the argument above about the asymmetric genetic relatedness predisposing species toward eusocial systems. A female is more related to a sister (0.75) than to her offspring (0.5); however, haplodiploidy also means that a female is less related to her brother (0.25) than to her offspring. If an equal number of females and males existed in the social group, the average relatedness of a female worker to her siblings would be (0.75 + 0.25)/2, or 0.5. Thus, there would be no particular reason to favor rearing siblings (brothers plus sisters) over rearing one's own offspring. But sex ratios in eusocial hymenoptera are not equal to 1; there are far more worker females than drone males. In honeybees, the queen controls the sex ratio through control of stored sperm. If an egg passes through the oviduct while sperm is released, a female results; if sperm is not released, a male drone is produced. By manipulating the sex ratio, the queen at the top of the hierarchy maintains a skew in the sex ratio that biases the workers toward sibling support rather than toward independent reproduction.

More direct behavioral coercion also occurs. Manipulating food deposited to developing larvae controls their size; small individuals become workers. Workers and in some cases the queen eat eggs produced by worker females. But most female workers never produce eggs, although they are capable of doing so. Pheromones produced by honeybee queens suppress the reproductive system in other females. Queens also produce queen mandibular protein (QMP), a pheromone that she feeds to attendant workers, which they, in turn, share with other bees in the hive. QMP has many functions, including attracting drones and, important here, inhibiting nursery workers from feeding royal jelly (the protein-rich nutrient necessary for producing the large queen) to developing larvae.

Queen honeybees only allow the production of new queens when conditions are right for swarming, the bee behavior in which a queen and a contingent of workers attracted to her pheromones leave the hive to establish a new nesting site. Before the swarm, queens lay fertilized (hence female) eggs in several large "queen cups." Workers provision these incipient new queens with royal jelly and cap the incubating larvae. Several new virgin queens eventually emerge to repopulate the hive. Even here, however, these closely related royal siblings seek each other out (most likely based on chemical signals) and attempt to kill each other. Unlike workers and soldiers, whose barbed stingers rip from the female after an attack and cause her death, queens are equipped with specialized stingers without barbs, which allow them to sting repeatedly. In some species of bees, continual conflict is apparent. Various species of stingless bees in the genus *Melipona* are eusocial and, like honeybees, have a single queen. But in the apparent absence of pheromonal coercion, they continually raise many competing queens, which are then attacked and executed when found.

CONCLUSIONS

Reproduction in the extended sense, conceptualized as inclusive fitness, lies behind the evolution and the functions of all bonding, affiliation, cooperation, and altruism. It also may relate mechanistically to more complex types of cooperative behavior, as the peptides that regulate familial bonding are now being recognized as contributing to affiliative behavior, social tolerance, and trust in broader contexts. Eusociality is a rare and extreme form of cooperative social behavior, but is a microcosm of the issues relevant to all prosocial interactions. Hidden behind the cooperative behavior are conflicts that must be addressed. In a sense, social bonding and affiliation, and social conflict and aggression, are linked, as both are driven by reproduction. Conflict and aggression will be taken up in the next chapter.

BIBLIOGRAPHY

Antala T, Ohtsukib H, Wakeleyd J, Taylore PD, Nowak MA. 2009. Evolution of cooperation by phenotypic similarity. *Proc Natl Acad Sci* **106:** 8597–8600.

Ball G F, Balthazart J. 2002. Neuroendocrine mechanisms regulating reproductive cycles and reproductive behavior in birds. In *Hormones, brain and behavior* (ed. Pfaff DW, et al.), Vol. 2, pp. 649–798. Academic, San Diego.

Carter CS, Keverne EB. 2002. The neurobiology of social affiliation and pair bonding. In *Hormones, brain and behavior* (ed. Pfaff DW, et al.), Vol. 1, pp. 299–337. Academic, San Diego.

Chao L. 1996. Evolution of polyandry in a communal breeding system. *Behav Ecol* **8:** 668–674.

Donaldson ZR, Young LJ. 2009. Oxytocin, vasopressin, and the neurogenetics of sociality. *Science* **322:** 900–904.

DuVal EH. 2007. Cooperative display and lekking behavior of the Lance-tailed Manakin (*Chiroxiphia lanceolata*). *Auk* **124:** 1168–1185.

Goodson JL, Schrock SE, Klatt JD, Kabelik D, Kingsbury MA. 2009. Mesotocin and nonapeptide receptors promote estrildid flocking behavior. *Science* **325:** 862–866.

Griffith SC, Owens IPF, Thuman KA. 2002. Extra-pair paternity in birds: A review of interspecific variation and adaptive function. *Mol Ecol* **11:** 2195–2212.

Haig D. 1997. The social gene. In *Behavioral ecology: An evolutionary approach*, 4th ed. (ed. Krebs JR, Davies NB), pp. 284–304. Blackwell, Oxford.

Hamilton WD. 1964. The genetic evolution of social behaviour. I and II. *J Theor Biol* **7:** 1–52.

Hammock EAD, Young LJ. 2005. Microsatellite instability generates diversity in brain and sociobehavioral traits. *Science* **308:** 1630–1634.

Hatchwell BJ, Komdeur J. 2000. Ecological constraints, life history traits and the evolution of cooperative breeding. *Anim Behav* **59:** 1079–1086.

Kirsch P, Esslinger C, Chen Q, Mier D, Lis S, Siddhanti S, Gruppe H, Mattay VS, Gallhofer B, Meyer-Lindenberg A. 2005. Oxytocin modulates neural circuitry for social cognition and fear in humans. *J Neurosci* **25:** 11489–11493.

Kosfeld M, Heinrichs M, Zak PJ, Fischbacher U, Fehr E. 2005. Oxytocin increases trust in humans. *Nature* **435:** 673–676.

Krakauer AH. 2005. Kin selection and cooperative courtship in wild turkeys. *Nature* **434:** 69–73.

Liu RC, Schreiner CE. 2007. Auditory cortical detection and discrimination correlates with communicative significance. *PLoS Biol* **5:** 1426–1439.

Nelson RJ. 1995. *An introduction to behavioral endocrinology.* Sinauer, Sunderland, MA.

Ptak SE, Lachmann M. 2003. On the evolution of polygyny: A theoretical examination of the polygyny threshold model. *Behav Ecol* **14:** 201–211.

Queller DC, Eleonora Ponte E, Bozzaro S, Strassmann JE. 2003. Single-gene greenbeard effects in the social amoeba *Dictyostelium discoideum*. *Science* **299:** 105–106.

Ratnieks FLW, Helantera H. 2009. The evolution of extreme altruism and inequality in insect societies. *Philos Trans R Soc Lond B Biol Sci* **364:** 3169–3179.

Regan CT. 1925. Dwarfed males parasitic on the females in oceanic Angler-Fishes (*Pediculati Ceratioidea*). *Philos Trans R Soc Lond B Biol Sci* **97:** 386–400.

Rosenblatt JS, Siegel HI, Mayer AD. 1979. Blood levels of progesterone, estradiol, and prolactin in pregnant rats. *Adv Study Behav* **10:** 225–311.

Ryan MJ, Rand W, Hurd PL, Phelps SM, Rand AS. 2003. Generalization in response to mate recognition signals. *Am Nat* **161:** 380–394.

Villinger J, Waldman B. 2008. Self-referent MHC type matching in frog tadpoles. *Proc R Soc Lond B Biol Sci* **275:** 1225–1230.

Walum H, Westberg L, Henningsson S, Neiderhiser JM, Reiss D, Igl W, Ganiban JM, Spotts EL, Pedersen NL, Eriksson E, Lichtenstein P. 2008. Genetic variation in the vasopressin receptor 1a gene (*AVPR1A*) associates with pair-bonding behavior in humans. *Proc Natl Acad Sci* **105:** 14153–14156.

Young LJ, Hammock EAD. 2007. On switches and knobs, microsatellites and monogamy. *Trends Genet* **23:** 209–212.

Young LJ, Insel TR. 2002. Hormones and parental behavior. In *Behavioral endocrinology* (ed. Becker JB, et al.), pp. 331–372. MIT Press, Cambridge, MA.

C H A P T E R 9

Conflict and Aggression

Eᴠᴇʀ ꜱɪɴᴄᴇ ᴅᴀʀᴡɪɴ ᴀʀᴛɪᴄᴜʟᴀᴛᴇᴅ ᴛʜᴇ ᴛʜᴇᴏʀʏ of evolution by natural selection, the-oretical biology has been based on conflict. Individuals compete for resources, including mates. Competitors may be other species (e.g., in predator–prey relation-ships), but more often than not they are other conspecifics. In some cases, the com-petition is for a resource that all individuals need: food, breeding territories, mates. In other cases, interacting individuals may be in conflict because they have different but competing interests. The male–female conflict related to mate choice described in Chapter 7 is the most important example.

Conflict can lead to aggression, particularly when it involves scarce resources. Animal social aggression can be violent and clearly aimed at injuring or killing an opponent. In many cases, however, social aggression manifests itself as a ritualized pattern of escalating responses and counterresponses that only reaches a physical attack if one party does not retreat. In these cases, species-typical displays serve as proxies for actual fighting, and animals have evolved signals that communicate their status as winners and losers.

ADULT SOCIAL AGGRESSION: DEFENDING SPACE AND RESOURCES

Chapter 8 made the point that social bonding, affiliative behavior, and cooperation are all related to reproduction. Social aggression is, behaviorally speaking, the oppo-site of those behaviors. But like those other behaviors, nearly all social aggression is related to reproduction. Access to mates, resources needed to attract mates, and defense of mates and offspring underlie social aggression. Just as for social bonding, social aggression is also tied mechanistically to reproduction. The hormonal and neural modulators of reproductive social behavior are the main drivers of social aggression.

Territoriality

It is important to distinguish a "territory" from a "home range." A home range is a geographic area that an individual uses habitually, often to forage for food. Individual home ranges can overlap and are not defended, although individuals may act aggressively if they happen to encounter each other. A territory is, instead, a smaller geographical area containing a resource of some kind that is actively defended against conspecific intruders. That resources may be a nest or other type of breeding site, which is the more common situation, or an area containing food or other resource needed to sustain the individual and its mates. An example of the latter is the territory defended by red deer (*Cervus elaphus*). In this species, males defend large areas of grassland on which they and their mates graze. Red deer have become a prime ethological model of mammalian social behavior. Their territorial behavior serves to illustrate basic features of animal territoriality and aggression and their link to reproduction, sex differences, and seasonality.

Red deer aggression is seasonal and tied closely to reproductive cycles (Fig. 9.1). The mating season for red deer begins in late summer and continues into the fall. Outside of the mating season, male deer live together in groups separate from female groups. Males within the groups are not very aggressive to each other. If anything, as in flocking birds and shoaling fish, individuals within such herds prefer to affiliate with conspecifics than to be alone. In late summer, male morphology and behavior change. Antlers, which have been growing throughout the summer, lose their soft outer covering and harden. Males bellow to advertise their location and reproductive

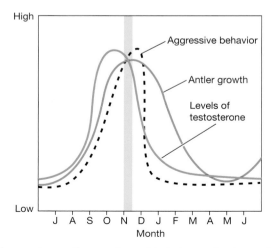

Figure 9.1. Levels of aggression, antler growth, and testosterone levels vary together seasonally in red deer. Gold bar indicates time of peak mating.

status. They become intolerant of other males, move to rutting areas, and fight each other to establish and defend territories. Males use the hardened antlers as weapons, butting and locking antlers, attempting to twist each other to the ground. The fights are serious and can result in injury and, rarely, death. The territories they fight over are areas of grassland onto which they attempt to attract females and from which they repel any intruding male. The size and quality of the territory are essential for successful mating. It must be large enough with sufficient grazing opportunities to maintain the presence of multiple females. If it is too small, females will wander away while they graze into the adjacent territory of another male. The larger the territory, the greater is the number of females that a male can maintain within it, but only at the cost of continually patrolling the territory to keep females in and other males out. The resource being fought over is technically a plot of land and the grass on it, but in reality the resource is the opportunity to mate.

Selection on the male for large body size, large antlers, and high levels of aggression is intense because all three contribute to his ability to win fights and control this resource. The selection is intensified by the fact that females come into estrus (i.e., become fertile and receptive) for only ~3 wk in October; missing that window means that a male misses an entire year of reproduction. Note that female red deer are under no similar selection pressure. As a result, there is a noticeable sex dimorphism in body size (males are significantly larger than are females), in body adornment and displays (females lack antlers and an area of thickened fur on the chest; males display vocally, females do not), and in behavior (males are more aggressive, and, although females will fight, they do so briefly using their front legs, a pattern males show only during the nonbreeding season). This is a classic case of polygynous mating (see Chapter 8) and the sex dimorphisms and male behavioral traits that emerge with it. As is typical for polygynous systems, no paternal care of offspring occurs. Females remain pregnant over the winter and give birth in the spring; offspring stay with the females in their foraging groups, whereas males stay separate. During this time, male aggressive territoriality wanes starting in late fall, and by midwinter the males have shed their antlers.

Not all territorial aggression is linked to polygynous mating. Many pair-bonded animals, such as birds, defend nest-based territories once mating has occurred. Here direct defense of the offspring and the resources that support them is the goal of territoriality, and both males and females can be aggressively territorial. As in the example of red deer, this territoriality is tied to the breeding season. It is not uncommon for birds to express highly territorial behavior during nesting, when they are intolerant of both conspecifics and any other birds that approach the nest, but then to aggregate in large communal foraging flocks in the nonbreeding winter months.

Overt fighting is common during the initial turmoil when territories are being established. Thereafter, ritualized displays often substitute. Such behavior leads to the evolution of ornamentation or behavioral patterns used for threat displays or advertisement of territorial status. As males are usually the more aggressive, males tend to be more ornamented than females are, either with visual markings, vocalizations, stereotyped behavioral displays, or other characters that are used to advertise the presence of a male on his territory. The substitution of a display for fighting is explained based on a benefit to both parties in a dispute. Stronger animals benefit by avoiding continual challenges from opponents, and weaker animals benefit by avoiding potentially injurious fights they would lose anyway. Both thereby save time and energy. Many mammals go a step further in their territorial signaling by marking territories chemically. Canines deposit scent cues through urination and defecation; cats and rodents by rubbing face and flank areas containing special sebaceous glands; and many ungulates have special glands on the face that they use to deposit pheromone droplets on plants or other substrates. The chemical signals used for territorial markings are relatively long-lasting, low-volatility substances that are often lipid based. In this way, a territorial resident can "display" his presence even when absent. In previous chapters, we noted that sex dimorphism in ornamentation was linked to sexual selection due to female mate choice. These are not mutually exclusive explanations; the same signals are often used for both territorial displays and for mate attraction as both are functionally linked.

Mate Guarding

When territories are mating sites as in red deer or in pair-bonded birds, males spend a considerable amount of time accompanying females and repelling approaching males. Even in nonterritorial animals, male mate guarding is commonly seen as males try to monopolize females when they are in a fertile and receptive condition. Mate guarding benefits the male by reducing opportunities for other males to fertilize the female by reducing extra-pair copulations and overwhelming them through high rates of insemination. Guarding can be either precopulatory or postcopulatory. The reason for the former is obvious (maintaining readiness to mate as soon as the female is receptive while dissuading potential rivals); however, guarding after copulation also benefits the male in some reproductive systems. If females store sperm, then preventing other males from inseminating the female after a successful copulation preserves the male's sperm investment, and continuing to copulate helps to overcome any extra-pair copulation that may occur (an example of sperm competition). Many invertebrates such as field crickets (*Teleogryllus* spp.) engage in postcopulatory mate guarding after the male transfers the spermatophore to the female. Sperm transfer from the spermatophore continues for some time after it is attached. During the postcopulatory mate guarding, male crickets actively interfere with female attempts to remove the spermatophore. Postcopulatory mate guarding in crickets therefore maximizes male sperm investment in two ways, by preventing the female from removing the spermatophore before all sperm has been transferred and by preventing rival males from gaining access to the female if she does remove it.

Leks and Display Sites

As discussed in Chapter 7, males of several species of birds, many frogs, and a few mammals and insects gather at communal sites and produce mating displays in leks. A lek is not a territory, because the lek as a whole is not defended and males do not try to exclude others from the group. Nor is it a resource-based area, because leks generally do not contain materials of value to individuals at the lek, and nesting does not occur on the lek itself. It is, instead, a site for competitive displays that function to attract females. Within the lek, individual males do maintain a personal space or distance from other males and will actively repel or otherwise act aggressively toward a male that gets too close. In essence, they try to maintain a personal territory. Display sites are often stable, and do vary in desirability. For example, in sage grouse, more central areas within the communal display ground are held by stronger, more aggressive individuals (Fig. 9.2). Moreover, their personal space (the area within which they will not tolerate another displaying male) is larger than that of smaller or younger males occupying the periphery.

The "Dear Enemy" Phenomenon

Once a territory is established, an individual holding that territory might encounter one of two types of conspecific rivals. One is the individual holding an adjacent territory. The other is a novel individual with no territory of its own. What are the costs and benefits of acting aggressively in each case? Seeing (or, in the case of a songbird,

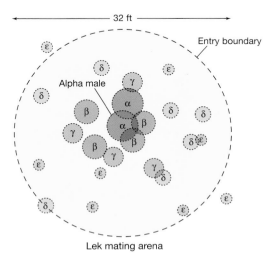

Figure 9.2. Representative lek organization in sage grouse. Greek letters indicate dominance organization, with α indicating the more dominant males. Dominant males occupy the center of the lek with progressively more subordinate males at the periphery.

hearing) the neighbor is a regular occurrence, so responding repeatedly would be energy intensive, and doing so constantly would subtract from time spent foraging or tending to partners and young. The benefit accrued by these regular fights or displays would be minimal: Each animal already holds a territory with established boundaries, so the result would most likely be the status quo. Encountering a novel intruder is different. The encounter is unpredictable: The capabilities and intent of the intruder are unknown. The risk of underresponding vastly outweighs the risk of overresponding because the potential outcome of not responding strongly could be displacement from the territory, or at least loss of territorial resources or mating success should the intruder succeed in copulating with the resident's partner. A rational approach to territorial aggression would be to decrease aggressive responses to neighbors with settled territories of their own while retaining vigorous responses to novel challenges. This is exactly what most territorial animals do. The phenomenon in which animals are less aggressive to familiar challengers than to novel ones is called the "dear enemy" effect.

The dear enemy phenomenon is a surprisingly sophisticated behavior. It requires individual recognition on the part of territorial animals. This, in turn, requires the animal to discriminate subtle differences in displays, to memorize the association of those signals with a location, and then to form expectations about where such a signal should and should not come from. Numerous playback experiments with songbirds show that territorial birds display all of these traits for the song of neighbors and, furthermore, that they respond differently when hearing a neighbor's song as opposed to a song from an unfamiliar bird. Remarkably, ant colonies also demonstrate the dear enemy effect. When workers from nearby colonies are placed together, they fight less than they do when workers from distant colonies are put together. At least one study on the ant *Acromyrmex lobicornis* showed that the differential behavior was not the result of the nearby ants being genetically related; the effect occurred despite a similar degree of genetic relatedness. All the available evidence suggests that the dear enemy effect is a learned one. Tom A. Langen and colleagues manipulated ant workers (from the species *Pheidole tucsonica* and *Pheidole gilvescens*) by allowing them to gain familiarity with ants from a different colony. They placed workers in a vial separated by a mesh divider from unfamiliar colony ants. This allowed them to receive odor cues, but not otherwise interact. Compared to a no-exposure condition, the exposure treatment significantly reduced fighting when the familiarized ants were later introduced to new ants from the colony to which they had been exposed (Fig. 9.3).

The dear enemy effect is also very sophisticated in its expression and closely tied to cost–benefit concerns across species and even within an individual. Ethan J. Temeles, in a classic review of the phenomenon, found that the effect is far more common in bird species that form mixed-purpose breeding and feeding territories than in species that defend breeding sites but forage elsewhere. Presumably, incursion on the multipurpose territory has potentially more costs. Individuals are capable of adjusting their responses depending on perceived threat. Jeremy Hyman found that

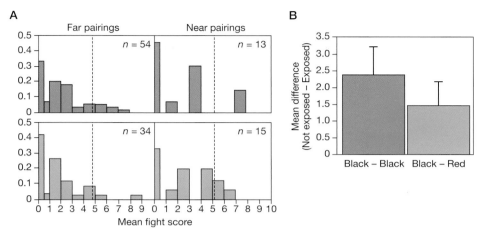

Figure 9.3. The "dear enemy" phenomenon in two species of ants. (A) Ants more often respond aggressively, and respond more aggressively when they do, to ants from colonies far (>6 m) away than near. "Fight score" values are scaled so that scores <5 are mild to nonaggressive social investigations, and scores >5 are aggressive social interactions. (B) Workers previously exposed to the odors of conspecific or heterospecific ants from other colonies respond less aggressively to ants from those colonies in later encounters. Mean + SE values of a difference score in fighting level (non-exposed minus exposed scores) is shown for conspecific (black–black) and heterospecific (black–red) pairing. No difference with experience would result in a difference score of 0.

the effect changed seasonally in Carolina wrens (*Thryothorus ludovicianus*). In the spring after territories had been stable, birds responded less to familiar songs than to novel songs. But in the fall, when additional birds enter the nesting territory, they responded equally to familiar and stranger songs. The neighbor's behavior also influences how vigorously a territorial bird responds to its song. Renee Goddard used playbacks to examine a resident's response to a simulated intrusion (the playback of a song in the resident's territory). In one condition in her study, the resident heard a playback of a neighbor's song from within the resident's territory, simulating an intrusion by the neighbor. Goddard found that after this manipulation, residents increased their singing in response to the neighbor's song compared to their levels before the intrusion. Here the resident was clearly responding to the perceived "cheating" by the neighbor by rejecting the dear enemy accommodation and treating the neighbor as a threat. A similarly sophisticated adjustment of territorial aggression based on potential risk is seen in pupfish (*Cyprinodon variegatus*). John K. Leiser found that pupfish males were less aggressive to familiar conspecifics than to unfamiliar conspecifics. However, when a female was present, the resident was equally aggressive to familiar and novel individuals (Fig. 9.4). This did not, however, change the male's aggressive response to heterospecifics, which remained constant regardless of condition or familiarity.

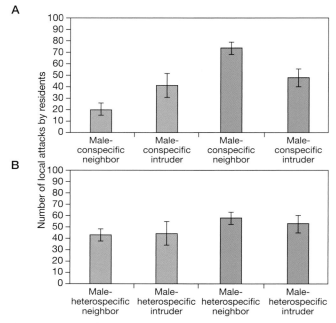

Figure 9.4. Number of times a male pupfish attacked an intruding male conspecific (*A*) or heterospecific (*B*) with (red bars) and without (blue bars) a female in its territory (mean ± SE). Males responded less to a neighbor than to an intruder when alone, but when a female was present, fighting increased to both, and the "dear enemy" tolerance of the neighbor was eliminated. There was little effect of a female's presence when the neighbor or intruder was a heterospecific.

This variability in the expression of the dear enemy effect shows more broadly that aggression is a carefully titrated behavior. It is sensitive to risk assessment and past experience, meaning that learning plays a significant role in its expression. These traits hold for other categories of aggression as well.

DOMINANTS, SUBORDINATES, AND SOCIAL HIERARCHIES

Many of the territorial organizations discussed above form in the context of dominance hierarchies. Dominance hierarchies are social orders in which individuals (often a single individual) at higher levels outcompete those at lower levels for resources (territories, food, mates) through aggressive social interactions. Dominance hierarchies are common whenever animals live in social groups and compete for limited resources. One of the defining features of dominance hierarchies of either sex is that the dominant individual has greater reproductive success through greater control of mating opportunities, either by mating with many more females (in the case of dominant males) or by mating with higher-ranking, and more resource-rich, males (in the case of dominant females). Most dominance hierarchy studies focus on males, where the phenomenon is more common (perhaps in part because competition for females by males is more intense than vice versa). In many primates, however, separate male and female dominance hierarchies coexist within a troop, with complicated interactions between them. High-ranking males interact with both high- and low-ranking females, for example, but low-ranking males risk serious injury by approaching a high-ranking female.

Dominance is defined behaviorally, both by an individual's own behavior and by the way in which other individuals react to it. An individual is said to be dominant if it acquires and can maintain a higher-quality territory, obtains more matings, or has primacy in the use of other resources (like food), and if other individuals defer to it. These lower-ranking, or submissive, individuals are easily displaced by dominant individuals, are often harassed by them, and back down from confrontations with dominants or actively avoid them. Dominant–subordinate relationships need to be established behaviorally through aggressive interactions, either direct fighting or escalating threat displays until one of the contestants gives up. Once established in a social group, relationships are relatively stable and require little overt fighting to maintain. Instead, aggressive displays by the dominant, exchanged with submissive displays by the subordinate, are sufficient. When this stage is reached, dominance hierarchies are remarkably good at reducing the level of aggression, and the risks it entails, for all members in the social group. Dominance hierarchies are rarely, if ever, strictly linear in nature. Instead, the most common pattern is more like a pyramid, with a dominant individual at the top, and progressively larger groups of individuals at each lower level in the dominance pyramid (see Fig. 9.2). Individuals are dominant to individuals at all levels below them, but not to those in levels above. Within levels, a variety of relationships might emerge that are not always strictly transitive.

Although dominance hierarchies are at some level based on dyadic interactions, at least in the initial stage, more complicated assessment strategies are also used. Social groups provide ample opportunities for individuals to observe agonistic interactions. Several studies of such bystander effects have been done in the Siamese fighting fish (Betta splendens). Male Bettas are territorial and aggressive and display to each other as in many other species. Several clever laboratory studies used one-way mirrors and other manipulations to allow a male to watch aggressive interactions between other male pairs without interacting with them or fighting itself. Such studies have consistently shown that fish use information about the outcome of the observed fight to inform their subsequent behavior. In one such study by Rui Oliveira and colleagues, bystander fish that were later paired with the fish they observed took significantly longer to approach and longer to display to the observed winners than to the observed losers (Fig. 9.5). Control experiments showed that this was not due to differential behavior in winners and losers. When paired with winners and losers of a contest they did not observe, the bystander fish approached and behaved equally to both.

A

B

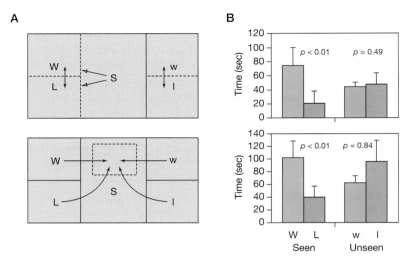

Figure 9.5. Male fighting fish (*Betta splendens*) adjust their aggression after watching the outcome of other fights. (*A*) Experimental apparatus and design. (*Top*) A test fish (S) in the center compartment can watch two fish on the left fight until one wins (W) and one loses (L), but they cannot see him; on the right, two other fish fight until one wins (w) and one loses (l) in a contest the test fish cannot see. (*Bottom*) Each stimulus fish is then paired with the test fish. (*B*) Latency of the test fish to approach (*top*) and display (*bottom*) to each of the stimulus fish: The stimulus fish acts differently only to perceived winners and losers.

Observational learning of this type has been seen in mammals (including primates) and birds as well.

Experience Determines Winners and Losers

Larger size (often coinciding with greater maturity) and physiological vigor certainly play a role in predicting that an individual will emerge from a contest as the winner. Context and experience play an even greater role. Individual primates, for example, can express either dominant or submissive behavior depending on whom they encounter within the social hierarchy. Furthermore, in most primate societies, juvenile males leave one troop to join another. When they do, even the more dominant animals adopt subordinate roles in the new troop and must gradually move upward though successive agonistic encounters. Both examples show that the behavior is not a fixed trait of the individual. The location of an encounter is also critical. It is universally the case that an individual on his home territory has a far greater chance of winning an aggressive interaction than does an intruder, even when the intruder is larger. Last, experience with aggressive social behavior is an important determinate of how an animal fares in future contests. Winners tend to win in the future; and a far stronger effect is that losers of targeted social aggression profoundly change their aggressive behavior and display persistent subordinate behavior.

Kim Huhman examined this "conditioned defeat" behavior in Syrian hamsters (*Mesocricetus auratus*). Hamsters are aggressive and very territorial. Placing a male hamster in the home cage of a larger male invariably makes the introduced male the target of rapid and vigorous territorial aggression, leading to that animal displaying submissive behavior and fleeing from the resident's continued approaches and attacks. Thereafter, the defeated animal seems incapable of shedding its subordinate status. Huhman tested the persistence of this state by removing the defeated animal from the larger male's cage and housing it separately in its own cage. Every 3–5 d a smaller, nonaggressive hamster was placed in the defeated animal's new home cage. Normal behavior would be for the resident to attack the intruder and dominate it. Instead, after that single social defeat, the hamster actively avoided the relatively passive intruder and fled from its nonthreatening exploratory approaches (Fig. 9.6). In one study, the submissive behavior persisted for 33 d, even in the absence of any additional aggression directed toward the defeated animal. Defeated hamsters also showed many physiological changes, including chronically high levels of stress hormones and greater stress reactivity, low levels of testosterone, and impaired immune responses. Hamsters may be extreme in their response to social defeat, but they are not unique.

Much of the work on experience and aggression has focused on adults. Yvon Delville's work provides an important counterpoint to the adult work by investigating the effect of aggression in prepubertal juvenile hamsters. Like many mammals, juvenile hamsters and other rodents engage in "play" fighting. An ethological analysis shows that as hamsters move from the prepubertal stage to reproductively mature

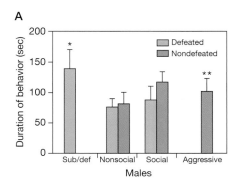
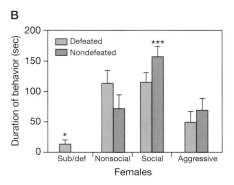

Figure 9.6. Defeated and undefeated hamsters respond differently in later social encounters with a nonaggressive conspecific, with some difference in the male (A) and female (B) responses. Defeated animals display more submissive ("sub/def") displays and fewer aggressive ("aggressive") displays, with the effect more pronounced in males (all mean+SE); differences between defeated and undefeated groups: (*) $p < 0.001$; (**) $p < 0.01$; (***) $p < 0.05$. Defeated females also show fewer overall social exploration behaviors ("social"). There were no differences in nonsocial behaviors such as grooming, sleeping, and feeding ("nonsocial").

adults, fighting patterns gradually move from a juvenile pattern (boxing-like attacks targeted to the front with little or no damage done) to an adult pattern (attacks targeted to the flank and rear, with a greater likelihood of drawing blood). Delville and Joel C. Wommack staged agonistic social encounters between smaller juveniles and larger aggressive adults in a paradigm similar to Huhman's social defeat, although with repeated rather than one-time interactions. Whereas Huhman showed that in adults social defeat results in persistent submissive behavior, Delville showed that when it occurred during the critical pubertal period, a much more complex behavioral response ensued. Juvenile hamsters targeted by aggressors accelerated their progression from juvenile to adult fighting patterns (Fig. 9.7). Furthermore, the pubertal defeats resulted in persistent changes in their later adult aggressive responses. Although defeated animals later showed normal-to-exaggerated submissive behavior when faced with larger opponents, they were far more aggressive than were control animals when encountering smaller or less aggressive opponents. Delville's work

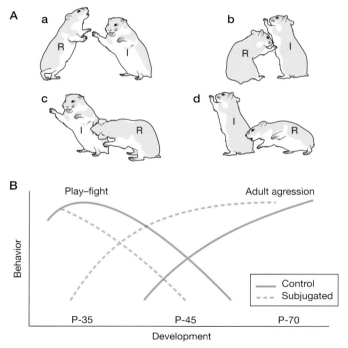

Figure 9.7. Development of aggression in hamsters. (A) Fighting takes several stereotyped forms and targets: (a) frontal play fighting, (b) flank attack, (c) lower belly, and (d) rump. The a and b forms are found in juveniles; the c and d forms are found in serious adult aggression. R, resident; I, intruder. (Bottom) Summary of developmental patterns from juvenile to adulthood in control animals and animals defeated repeatedly during the juvenile period. Defeated (subjugated) animals lose juvenile play fighting faster and gain adult aggression sooner.

provides an important developmental perspective on aggression by showing that behavioral interactions early in adulthood are as important as adult experience in setting an individual's pattern of social aggression.

NEURAL AND HORMONAL MECHANISMS OF AGGRESSION

On average, aggression is higher in males than in females; male aggression rises at the onset of sexual maturity (puberty); and in seasonally breeding animals, male aggression levels along with specialized behavior and morphology linked to agonistic displays track seasonal reproductive behavior. All three factors point to the gonadal steroid hormones as key regulators of social aggression, as well as the functional link between aggression and reproduction. Gonadal hormones, in turn, interact with two brain neuromodulators. One is vasopressin/vasotocin (AVP/AVT), one of the nonapeptides important in modulating reproductive social behavior (see Chapter 8). The other is the monoamine neurotransmitter serotonin, which acts as an important neuromodulator of many brain functions, including emotional responding and mood. Serotonin plays an important regulatory role in aggression in both vertebrates and invertebrates, despite their very different brain structures.

Male Social Aggression

Numerous castration-implant studies and other manipulations show that, in males, the presence of testosterone during a critical pre- or perinatal developmental period is necessary to organize the brain for adult male social aggression. Thereafter, adequate levels of testosterone that rise with the onset of reproductive maturity are necessary for the expression of male social aggression. This dual "organizational–activational" role of gonadal steroids was explained in Chapter 6. One example given there was the classic work on how intrauterine position affected later adult behavior: Female rat pups that develop between two males, and are thus exposed to their leaking fetal androgens, are more aggressive than pups that grow between two females. Experimental manipulation of developmental androgens directly confirms this. In adult animals, one of the most consistent experimental findings in behavioral endocrinology is that removing androgens by castration reduces aggression. Even more dramatically, castration reduces the agonistic displays males produce to advertise their positions on territories or display sites to males and to attract females, and subsequent implantation of testosterone restores this behavior (Fig. 9.8).

Testosterone is converted into various metabolites in its target tissues (including the brain), with the two most important being another androgen, dihydrotestosterone (DHT), and estrogen, the principle hormone released by the female ovaries (see Chapter 6). Oddly enough, brain masculinization occurs when neurons take up and convert testosterone into estrogen intracellularly, and then estrogen binds to

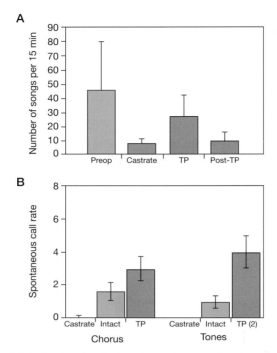

Figure 9.8. (A) Territorial singing in zebra finches drops after castration, is restored by implanting testosterone propionate (TP), and then drops again if the testosterone implant is removed. (B) Similarly in frogs, spontaneous calling is absent in castrated males, compared to normal males, and testosterone implants raising levels above normal average results in high levels of calling. The effects are present whether or not frogs had previously been exposed to conspecific calls (Chorus) or to control random tones (Tones).

intracellular estrogen receptors to act as a transcription factor. There remains some controversy as to whether androgens (testosterone or DHT) or converted estrogen activate the adult expression of aggression and agonistic signaling. The best current guess is that both metabolic pathways are important and that the balance between these two interrelated routes may vary across species. How androgens affect the brain through either pathway to then affect behavior is complex and not completely understood. It is clear, however, that at least part of the mechanism is the influence of androgens on the actions of two brain neuromodulators, AVP/AVT and serotonin.

AVP/AVT was discussed in Chapter 8 as a modulator of male social bonding and paternal behavior, acting in the brain through the V1a receptor. This same peptide, using the same V1a receptor, appears equally important in facilitating male aggression and the agonistic signaling that goes with it. In hamsters, AVP is released into the anterior hypothalamus when a male encounters an intruder. Experimentally infusing AVP there (or as in most experiments giving it systemically) increases aggression in hamsters to appropriate targets like an intruder. Conversely, blocking V1a receptors in this region blocks attacks against an intruder. Experimental manipulations of AVP and testosterone in various combinations show that AVP's effect on aggression depends on the presence of testosterone. Directly infusing AVP into the anterior hypothalamus does not activate attacks or other forms of aggression in castrated hamsters, but does so when castrated hamsters are given replacement testosterone. Testosterone treatment also increases the levels of AVP in the brain. Testosterone thus appears to act on AVP in two ways, by increasing its availability and by facilitating the capacity of V1a receptors to receive and/or transduce it into a neural response.

Although the broad strokes of AVP/AVT's actions are clear—this peptide can facilitate male aggression along with other male social behaviors—the details and species variability of its actions rapidly become complex. In several territorial species of birds and fishes, elevating AVT decreases aggression, whereas in congeneric gregarious, colonial, or otherwise nonterritorial species, the same treatment increases aggression. There is some evidence that different sites of action in the brain have opposite effects. In the lateral septum, AVP/AVT elevation may inhibit aggression, the opposite to what occurs with elevation in the anterior hypothalamus. It should be noted that species differences among voles in AVP effects on social attachment are now well established and appear to stem from differences in receptor density or distribution (see Chapter 8). The same may be true for AVP/AVT variation with regard to aggression.

Whereas AVP/AVT generally facilitates male aggression and agonistic signaling (with notable exceptions across vertebrate species), serotonin reduces aggression in vertebrates. There are an extraordinarily large number of pharmacologically distinct serotonin receptors in the brain; serotonin acts selectively through the 5-hydroxytryptamine 1 (5-HT1) class of receptors in influencing aggression. How gonadal steroids interact with serotonin signaling is not well understood. It appears as though the steroid hormones do not change the production or availability of serotonin itself. Rather, androgens, most likely acting though their conversion to estrogen, seem to act at the location of serotonin's receptors, most strongly in a hypothalamic area termed the medial preoptic area. Serotonin and AVP also interact in these brain areas. AVP containing neurons have 5-HT1 receptors on them, and activation of these receptors by a serotonin agonist blocks the normally facilitative effect of AVP on aggression. As with AVP/AVT, there are species differences in serotonin effects. One of the most interesting is that serotonin has universally been reported to have an inhibitory effect on aggression in vertebrates, but in invertebrates the opposite is often true: Elevated serotonin increases aggression in crayfish, for example. In ants, serotonin has context-dependent effects. It decreases aggression toward conspecific intruders but raises aggression against heterospecific intruders.

Although there are significant and interesting variations on this general pattern, there are some general mechanistic starting points for male aggression (Fig. 9.9): (1) Testosterone is important in vertebrates both developmentally and in adults in facilitating aggression; (2) AVP/AVT influences aggression and social signaling, usually facilitating

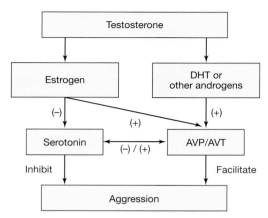

Figure 9.9. Diagram of the most commonly reported effects and interactions of testosterone, serotonin, and vasopressin/vasotocin (AVP/AVT) on aggression in vertebrates. (+) indicates an increase and (−) a decrease on the target. In general, serotonin inhibits and AVT/AVP facilitates aggression in rodents, but species vary considerably in serotonin and AVP/AVT effects on aggression.

but sometimes inhibiting it depending on the species; and (3) serotonin is an important regulator of aggression in both vertebrates and invertebrates and is most often inhibitory to its expression. It is important to note that these factors are only part of a far more complicated regulatory network involving many different neurochemical and endocrinological systems that influence social aggression and other social behaviors.

Female Social Aggression

Understanding the mechanisms of female aggression is complicated because the level of female aggression varies greatly across species, and within species it is expressed differently in different contexts. Female vertebrates produce a considerable amount of testosterone, and although much of it is converted to estrogen in the ovaries before secretion into the circulatory system, females often do have significant levels of circulating testosterone that may be equal to male levels. As in males, testosterone does seem to facilitate female aggression in some circumstances. The most consistent finding is that elevated testosterone causes increases in female aggression toward female intruders, particularly in birds when the females compete for nesting sites or males. Many studies, however, have found no consistent relationship between androgens and female aggression. Two hormones important for female reproductive behavior, progesterone and estrogen, have also been shown to influence female social aggression. In mammals, for example, maternal aggression is typified by a sudden increase in highly intense attacks toward intruders coming near nesting sites or infants. The behavior arises at or near birth. This is the same time when other maternal behaviors, such as nursing, huddling, and pup retrieval emerge (Chapter 8). Peaks in estrogen around birth are essential for those behaviors, and increased postpartum estrogen increases maternal aggression. Conversely, progesterone has often been found to inhibit female aggression in rodents as well as in some birds. In most mammals, progesterone is high during female receptivity and pregnancy; it then rapidly declines near birth, as estrogen rapidly rises. This reciprocal change in the two hormones is consistent with the different aggressive profiles of females pre- and postpartum: prepartum, during the behavioral phase when females are receptive toward males, estrogen, an aggression facilitator, is low, and progesterone, an aggression inhibitor, is high. Postpartum, when females actively and very aggressively defend nest sites and offspring from both male and female unfamiliar intruders (and often familiar males in non-pair-bonding species), facilitatory estrogen is high, and inhibitory progesterone is low.

The Testosterone "Surge" after Winning

A consistent finding from numerous vertebrate taxa is that winning an aggressive encounter triggers a fast, large, but relatively brief rise in circulating testosterone. This testosterone surge has often been found in males and has even been reported in some female birds after successful aggression triggered by a nest intruder. The surge typically occurs within minutes of the end of a successful aggressive interaction and lasts for minutes to at most a few hours. After the surge, the animal's baseline levels of testosterone return to prefight conditions. Thus there is no persistent endocrinological consequence of the surge.

Because the testosterone surge occurs after a behavior it cannot influence that behavior, and given that the elevation does not persist, its function is not to allow behavior to reset the hormone to a higher level. There is now a growing appreciation for the possibility that the testosterone surge is related to future behavior, not by changing baseline hormonal state but by acting on the brain to make aggression more likely. There are two possibilities. One is that testosterone facilitates neural plasticity and in so doing enhances an animal's ability to learn and adjust its behavioral response to future agonistic challenges. A second is that the testosterone surge acts (by triggering dopamine release) as a reinforcer—that is, as a "reward" that makes a similar response in a later, similar context more likely. In either case, the result would be that winning a fight would make an animal more likely to be aggressive when it finds itself again in a similar situation. There have been no definitive studies that show either to be the case for social aggression, but there is substantial support from basic neuroscience studies of brain plasticity and learning to support both possibilities.

LEARNING AND AGGRESSION

A recurring theme in this chapter's discussion of social aggression is that experience has a large influence. Stated another way, learning is an important mechanism for organizing an animal's aggressive behavior. Many changes in aggression with experience can be seen in terms of two simple learning processes: (1) habituation, a decrease in responding with repeated experiences, something applicable to the dear enemy phenomenon; and (2) sensitization, an increase in responding with repeated experiences, something applicable to the observation that engaging in aggression leads to heightened aggression in future encounters. Habituation in learning theory is a specific response adjustment to repeatedly encountered signals, not a blanket decrement in behavior. It allows an individual to avoid expending unnecessary energy or encountering unnecessary risk resulting from constantly responding to stimuli that are repetitive and proven to be both unthreatening and unrewarding. As in the dear enemy phenomenon, once territories are established by a period of relatively intense agonistic interactions, territory holders gradually spend less time responding to the display signals of immediate and familiar neighbors; that is, they habituate to those particular signals. But as in classically defined habituation, territorial animals remain perfectly capable of vigorous aggression or agonistic displays to a novel signal or even a familiar signal that comes from an unexpected location. Sensitization is the phenomenological opposite of habituation. Repeated exposure to a stimulus primes the individual to respond more vigorously to successive presentations of that stimulus.

Eun-Jin Yang demonstrated how both habituation and sensitization can occur by using video playbacks to green anole lizards (*Anolis carolinensis*). Male anoles respond to other males with stereotyped displays including unfolding and displaying the dewlap (a skin flap under the jaw), head bobs and forearm pushups, and a lateral compression display in which they turn their flank toward an opponent and compress their body to enlarge their profile. Yang presented male anoles in their home cage with the same video of a displaying conspecific or a control video of moving green objects each day for 6 d. The video clip showed the anole jumping off its perch at the end of 10 min, which is what a losing male might do. Males initially produced agonistic displays to the conspecific video, but little to the control video. Males seeing the conspecific video (but not the control males) increased responding each of the first 3 d (sensitization), but then gradually decreased until they were not responding to the conspecific video any more than were the control animals to their videos (habituation) (Fig. 9.10). Yang then introduced a novel male intruder into the home cages of experimental and control animals. Despite the fact that males that had viewed and responded to the video had decreased their aggressive responses down to that of the control animals, they responded significantly faster and more vigorously to the intruder. Yang's results thus revealed a third element to social learning: Prior experience with an aggressive challenge made the animals more aggressive in future social encounters when faced with a new challenge.

One might think that the physical and emotional stress, not to mention the potential or real pain of a fight, would render aggression aversive, or in the parlance of learning theory, would act as a negative reinforcer. Certainly, losing a fight is aversive, as indicated by the conditioned defeat phenomenon discussed above. Many naturalistic

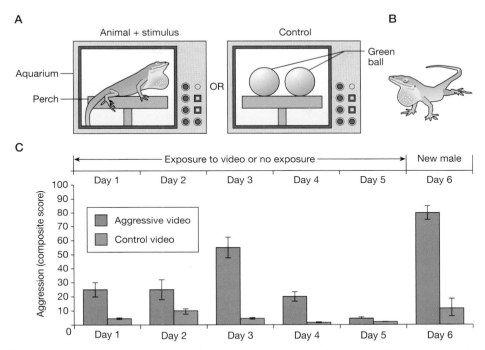

Figure 9.10. Experimental procedure for video playback test of repeated simulated aggressive encounters (A) and resulting aggression levels in *Anolis carolinensis* (B). Anoles first become sensitized to the video display, and then they habituate to it. However, this experience makes them more aggressive (compared to animals watching a control video) when they encounter a new male. (C) Statistical composite score of all aggressive display indices in male green anole lizards.

observations and experimental learning studies, however, have shown that successful aggression is positively reinforcing. Individuals that behave aggressively and win (or at least do not lose), like the anoles above, act more aggressively toward later challengers. That is, the experience has made that particular behavioral response, out of the many that are possible, more likely in the future. Many studies in diverse organisms have found that engaging in aggression (without losing) is rewarding. Territorial fish, for example, will work by swimming through obstacles or pushing a trigger to gain access to a conspecific to which they can make agonistic displays.

A direct test of the rewarding properties of a stimulus is called the conditioned place preference. In this learning paradigm, a subject is provided with a stimulus or event in one of two clearly distinct test chambers. After several pairings of the stimulus with the particular place, the subject is given the choice to be in either of the two environments when the stimulus is absent. If the event was aversive, the subject will choose the chamber where it never occurred; if, on the other hand, the event was rewarding, the subject will choose to spend time in the place where it had occurred. That is, the animal was conditioned to prefer that particular place because it was associated with an event it found rewarding. William J. Farrell used a conditioned place preference procedure to find out if aggression was rewarding in the *Anolis* lizards like those used in the learning study presented above. In one part of a two-part experimental chamber with different visual and textural features, male anoles were exposed to a mirror (Fig. 9.11). Anoles and many other non-mammals will display to their own image in the mirror, which, of course, responds in kind. After 10 min, the mirror was removed. Following 8 d of this, the anoles were allowed to move freely between the two experimental chambers. Males that had displayed aggressively toward the mirror demonstrated a clear preference for being in the chamber where they had previously engaged their image. Males that were presented with the mirror but never displayed to it and control males that spent equal time in the test chamber

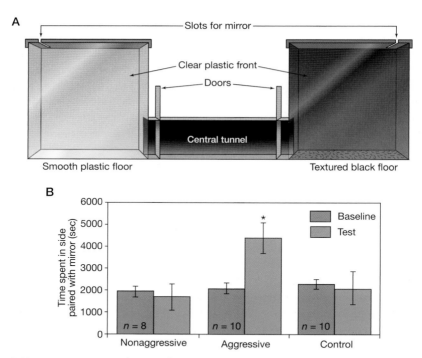

Figure 9.11. (A) Apparatus used to test for a conditioned place preference in green anole lizards. Males are exposed to a mirror in one distinct compartment but not the other. (B) After repeated pairings, males that responded with aggressive displays spent more time in the compartment where they previously saw the mirror, whereas control animals and animals that did not act aggressively spent equal time in the two compartments (mean ± SE time spent in the chamber where the mirror had been presented, measured during the test period with no mirror visible). The asterisk indicates a significant difference between baseline (before the mirror experience) and test (after the mirror experience) in aggressive animals.

without a mirror spent equal time in the two different chambers. Farrell's experiment showed that something about engaging in agonistic display bouts was rewarding to lizards.

Conditioned place preferences supported by aggressive encounters have been demonstrated in mice and hamsters as well. Anoles, like rodents and many other vertebrates, show a testosterone surge following a successful aggressive interaction. As discussed above, it has been speculated that this surge, acting on brain dopamine systems, is the mechanism underlying reinforcement learning such as this. Definitive experimental evidence of this has not yet been obtained.

INFANTICIDE AND SIBLICIDE

In Chapter 8 we stressed how the dictates of reproductive success lead to behaviors facilitating the survival of offspring. But nature being nature, reproductive success can dictate the opposite. Infants are often the target of aggression and killed. When adults are the aggressor, they nearly always direct attacks toward unrelated conspecific young with the intention of killing them, the phenomenon of infanticide. Sometimes, however, close relatives are the aggressor, with the most common cases being aggression among juvenile siblings leading to the death of a nestmate. This phenomenon is termed *siblicide*. It is not clear if adults (parents) contribute directly to siblicidal attacks, but curiously it is universally found that they do little to prevent it.

Infanticide among animals is not a random occurrence. It is a strategic behavior by males to increase their reproductive success. Males preferentially target offspring that are not their own, then mate with the mothers of those offspring. In primates and other large mammals, infanticide is most often seen in single male mating systems when the dominant male responsible for the majority of the group's offspring is displaced by an intruder or otherwise is removed from the social group. The intruder male may then kill the former resident's offspring. One function of this behavior is that it removes resource competitors to a male's own subsequent offspring. But a second and likely more important and immediate function is to bring females back into a reproductive state. As discussed in Chapter 8, parental and reproductive behaviors are in conflict, and differing hormonal states play against each other. By removing the stimuli maintaining parental behavior (i.e., the infants), the males accelerate the shift to the next reproductive state (mating). The phenomenon that rodent female reproductive receptivity is triggered when infant pups are removed has been well known from laboratory experiments since the 1960s. In fact, in mice just the chemical signals of a strange male can induce spontaneous pregnancy termination in females, a phenomenon called the Bruce effect (named after Hilda M. Bruce, who noticed this phenomenon in experiments performed with mice in 1959). Laboratory manipulations of receptivity and the Bruce effect are experimental observations from physiological psychology studies, but their relevance to the behavioral ecology of infanticide is obvious.

After loss of offspring or pregnancy termination, female rodents quickly resume reproductive cycling and receptivity, providing a clear reproductive benefit to the male aggressor.

Joseph Soltis and colleagues carefully documented an instance of infanticide in a wild population of Japanese macaques that contained multiple males and females. Although infanticide is rare in such social groups, and very rare in natural populations of macaques, the data are instructive for understanding the roots of infanticide behavior. The attacks by eight of the resident males occurred after high-ranking males had left the population, occurred early in the breeding season (when females would still be able to reproduce successfully), and selectively targeted unweaned infants (in macaques, lactation suppresses ovulation). Soltis used DNA analysis to show that the infants that were attacked were not sired by the male attackers. Last, males were on average eight times more likely to attack a female's infants if they had not previously mated with her. The dynamics of Japanese macaque populations support females mating with multiple males, as male residence in troops generally lasts only a few years, and in multiple male aggregations a single dominant male is incapable of monopolizing all female matings. The infanticide pattern in this particular instance suggests that promiscuous female behavior (which might be analogous to extra-pair copulations common in many species) could offer some protection from male infanticidal attacks due to the ambiguity in parentage that promiscuous behavior causes.

In a classic early review of infanticide, Sarah Blaffer Hrdy suggested that the threat of infanticide influenced the evolution of sexual strategies in female primates that make parentage ambiguous. These include the shift in higher primates away from strictly cyclic estrous receptivity to constant availability with receptivity determined by social interactions or situational context, and the use of behavioral signals of receptivity as opposed to morphological or chemical cues. Under these conditions, females are capable of courting and copulating with males regardless of their immediate fertility (including when they are already pregnant), and hence the true parentage of their offspring remains in question. An increase in the tendency of females to associate with males after giving birth is also apparent in many primates, the argument being here that it may offer protection from infanticidal attacks. It is admittedly difficult to prove that any of these primate traits evolved in response to the threat of infanticide. It is nevertheless important to think about how intraspecific aggression may have influenced the evolution of reproductive social strategies.

Although disturbing (to humans), infanticide is understandable from basic principles of evolutionary animal behavior. Siblicide is more puzzling. Here, one sibling targets another, often leading to the death of that closest of relatives. Siblicide is seen in a scattering of diverse bird taxa and some mammals. It can range from a facultative behavior seen only in the context of some nesting or environmental conditions, to an obligate behavior, meaning that it is the expected outcome of a shared nest. At one end, the aggression can reflect competition among nestmates or littermates for food, resulting in the death of one or more of the offspring that are outcompeted by their siblings. Such siblicide routinely follows a birth order hierarchy in which the first born or hatched is the aggressor toward the younger siblings or outcompetes them for food. At the extreme end, the older sibling simply kills the younger. The laughing kookaburra (*Dacelo novaeguineae*), an Australian kingfisher, generally lays three eggs asynchronously. Nestlings have a specialized hook on their beak that they use to attack each other. In about half of all nests, the youngest sibling dies as the older nestlings grab it by the head with the help of the juvenile hook and shake it violently; in some of the remaining nests, the youngest dies of starvation due to the constant harassment by its older siblings. Oddly enough, kookaburras are cooperative breeders: Mated pairs receive help provisioning their nestlings from older offspring that have fledged and grown to young adulthood. Thus, as the birds mature, they move from trying to kill each other to helping to raise their younger siblings.

This type of behavior seems difficult to reconcile with the basic evolutionary argument that increasing gene frequency is the goal of reproduction, as killing a sibling, which shares many of your own genes, in essence kills part of your extended reproductive fitness. This is the "inclusive fitness" that was articulated by W.D. Hamilton and has become a foundation for understanding natural selection and a variety of social behaviors that cannot easily be explained based on the value to the individual alone. Because of this apparent inconsistency, early accounts of siblicide attributed it to the unintended byproduct of high species aggression or pathological responses to abnormal conditions. But as the kookaburra example shows, the predictable nature of the behavior, the restriction of the targeted violence to a specific juvenile period, and, in this species, the presence of a specialized morphological adaptation to facilitate the aggression suggest something other than random pathology or overgeneralization of aggressive tendencies. Currently, siblicide is seen as an evolutionary strategy related to resource limits and parents' attempts to increase their own fitness in response to these limits.

It is likely the case that in all species expressing parental care, parents are capable of producing more fertile zygotes than they can possibly care for, and in many species regularly do. Female mammals such as pigs routinely have litters larger than the number of nipples available for nursing, and a variety of birds produce more eggs than the number of hatchlings they can comfortably provision. In both cases, this results in aggressive competition among the offspring for the limited resources available. One reason why parents do this has been explained by an "insurance hypothesis." Parents, especially avian parents that depend on foraging for prey to feed hungry nestlings, cannot predict from year to year how good—resource-rich or predator-free—any particular year will be. Yearly overproduction of young provides increased overall fitness to the parents at little additional cost, as in good years the "extra" offspring will survive, and it offers some insurance against one of the young dying for any reason. But the fitness benefit only exists if in bad years the extra offspring are quickly culled so that the remaining young survive to reproduce themselves, and the parents are not burdened with extra costs. Siblicide allows the offspring to "decide" among themselves whether culling is needed. This interpretation yields a prediction that can be tested: Aggression and siblicide should vary inversely with food availability.

The idea that resource constraints are proximate drivers of siblicide is called the "food amount hypothesis." There are alternative ideas linked to food quality as the primary driver, but these too are hypotheses about proximal cause linked to resources. Hugh Drummond reviewed the evidence supporting this in several families of birds where facultative siblicide is seen. In both instances of natural food deprivation and experimental restriction of food availability, nestling aggression increased. For example, Drummond reported that unmanipulated blue-footed booby nests in which the junior nestlings were killed had senior nestlings with a 20%–25% body weight deficit compared to nests where both nestlings survived. The hypothesis is that the discrepancy resulted from higher aggression by senior nestlings induced by a relative lack of food. Drummond and colleagues followed this natural observation experimentally by reducing the ability of nestlings to ingest food supplied by the parents by placing tape around the necks of nestmates. In control nests, tape was applied, but removed when the parents attempted to provision the nestlings with their typical mouth-to-mouth feeding. Natural sibling aggression occurred in both groups, but the food-restricted dominant (older) chicks progressively increased their aggressive pecking at their younger siblings, reaching a 349% increase by the third day. Removing the tape and allowing the food-restricted nestlings to eat normally caused a decline in their previously exaggerated aggression. Similar observations have been reported in spotted hyenas, one of the few mammals where facultative siblicide is seen. There, intersibling aggression is higher in lower-weight cubs, which occurs when mothers spend less time nursing. It may be that not all siblicide is directly related to food quantity, but the consistency in observational and experimental results across species suggests that it is a major determinate in many organisms.

CONCLUSION

Despite how humans may feel about it, aggression is a natural part of animal social behavior. Its function is to acquire and protect resources and to facilitate reproduction, and its evolution is driven by the same inclusive fitness considerations that underlie all social behavior. Given its tie to reproduction, it is not surprising that mechanisms of aggression are linked to mechanisms of reproduction. The same hormones and peptides that regulate mating and pair bonding regulate social aggression. In a real sense, the last three chapters in this section reflect different aspects of the same basic evolutionary drive—transferring genes from one generation to the next.

BIBLIOGRAPHY

Bruce HM. 1959. An exteroceptive block to pregnancy in the mouse. *Nature* **184:** 105. doi: 10.1038/184105a0.

Burmeister SS, Wilczynski W. 2001. Social context influences androgenic effects on calling in the green treefrog (*Hyla cinerea*). *Horm Behav* **40:** 550–558.

Clutton-Brock TH, Albon SD, Gibson RJ, Guinness FE. 1979. The logical stag: Adaptive aspects of fighting in red deer (*Cervus elaphus* L.). *Anim Behav* **27:** 211–225.

Cowlishaw G, Dunbar RIM. 1991. Dominance rank and mating success in male primates. *Anim Behav* **41:** 1045–1056.

Drummond H. 2001. A revaluation of the role of food in broodmate aggression. *Anim Behav* **61:** 517–526.

Farrell WJ, Wilczynski W. 2006. Aggressive experience alters place preference in green anole lizards (*Anolis carolinensis*). *Anim Behav* **71:** 1155–1164.

Ferris CF, Melloni RH Jr, Koppel G, Perry KW, Fuller RW, Delville Y. 1997. Vasopressin/serotonin interactions in the anterior hypothalamus control aggressive behavior in golden hamsters. *J Neurosci* **17:** 4331–4340.

Gleason ED, Fuxjager MJ, Oyegbile TO, Marler CA. 2009. Testosterone release and social context: When it occurs and why. *Front Neuroendocrinol* **30:** 460–469.

Godard R. 1993. Tit for tat among neighboring hooded warblers. *Behav Ecol Sociobiol* **33:** 45–50.

Hockham LR, Vahed K. 1997. The function of mate guarding in a field cricket (Orthoptera: Gryllidae; *Teleogryllus natalensis* Otte and Cade). *J Insect Behav* **10:** 247–256.

Hrdy SB. 1979. Infanticide among animals: A review, classification, and examination of the implications for the reproductive strategies of females. *Ethol Sociobiol* **1:** 13–40.

Huber R, Smith K, Delago A, Isaksson K, Kravitz EA. 1997. Serotonin and aggressive motivation in crustaceans: Altering the decision to retreat. *Proc Natl Acad Sci* **94:** 5939–5942.

Huhman KL, Solomon MB, Janicki M, Harmon AC, Lin SM, Israel JE, Jasnow AM. 2003. Conditioned defeat in male and female Syrian hamsters. *Horm Behav* **44:** 293–299.

Hyman J. 2005. Seasonal variation in response to neighbors and strangers by a territorial songbird. *Ethology* **111:** 951–961.

Kelley DB, Brenowitz E. 2002. Hormonal influences on courtship behavior. In *Behavioral endocrinology* (ed. Becker JB, et al.), pp. 289–330. MIT Press, Cambridge, MA.

Langen TA, Tripet F, Nonacs P. 2000. The red and the black: Habituation and the dear-enemy phenomenon in two desert *Pheidole* ants. *Behav Ecol Sociobiol* **48:** 285–292.

Legge S. 2002. Siblicide, starvation and nestling growth in the laughing kookaburra. *J Avian Biol* **33:** 159–166.

Leiser JK. 2003. When are neighbours 'dear enemies' and when are they not? The responses of territorial male variegated pupfish, *Cyprinodon variegatus*, to neighbours, strangers and heterospecifics. *Anim Behav* **65:** 453–462.

Mock DW, Parker GA. 1998. Siblicide, family conflict and the evolutionary limits of selfishness. *Anim Behav* **56:** 1–10.

Nelson RJ. 1995. *An introduction to behavioral endocrinology.* Sinauer, Sunderland, MA.

Oliveira RF, McGregor PK, Latruffe C. 1998. Know thine enemy: Fighting fish gather information from observing conspecific interactions. *Proc R Soc Lond B Biol Sci* **265:** 1045–1049.

Simon NG. 2002. Hormonal processes in the development and expression of aggressive behavior. In *Hormones, brain and behavior* (ed. Pfaff DW, et al.), Vol. 1, pp. 339–392. Academic Press, San Diego.

Soltis J, Thomsen R, Matsubayashi K, Takenaka O. 2000. Infanticide by resident males and female counter-strategies in wild Japanese macaques (*Macaca fuscata*). *Behav Ecol Sociobiol* **48:** 195–202.

Temeles EJ. 1994. The role of neighbours in territorial systems: When are they 'dear ememies'? *Anim Behav* **47:** 339–350.

Wommack JC, Taravosh-Lahn K, David JT, Delville Y. 2003. Repeated exposure to social stress alters the development of agonistic behavior in male golden hamsters. *Horm Behav* **43:** 229–236.

Yang E-J, Phelps SM, Crews D, Wilczynski W. 2001. The effects of social experience on aggressive behavior in the green anole lizard (*Anolis carolinensis*). *Ethology* **107:** 777–793.

Credits

Abbreviations: AAAS, American Association for the Advancement of Science; CDC, Centers for Disease Control; CSHLP, Cold Spring Harbor Laboratory Press; IEEE, Institute of Electrical and Electronics Engineers.

Chapter 1

1.1, top left, Based on Bradbury JS, Vehrencamp SL, *Principles of animal communication,* © 1998 Sinauer Associates; **1.1, top right,** reprinted from Nottebohm F, 2005, *PLoS Biol* 3: e164; **1.1, bottom right,** reprinted from Price JJ, Lanyon SM, *Evolution* 56: 1514–1529, © 2002 with permission from John Wiley & Sons Inc.; **1.3, bottom,** courtesy of Alexandra Basolo; **1.4,** reprinted from Chittka L, Doring TF, 2007, *PLoS Biol* 5: 1640–1644; **1.5,** reprinted from Sapolsky RM, *Nature Neurosci* 7: 791–791, © 2004 with permission from Macmillan Publishers Ltd.

Chapter 2

2.1, Reprinted from Orr HA, 2009, *Sci Am* 300: 44–51, © Tommy Moorman; **2.2,** reprinted from Smith JM, *Evolution and the theory of games,* © 1982 with permission from Cambridge University Press; **2.3,** redrawn with permission from Sherman PW, 1980, *Sociobiology: Beyond nature/nurture?* (eds. Barlow GW, Silverberg J); **2.4,** © Alex Wild/Visuals Unlimited; **2.5,** reprinted from Hurst GDD, Werren JH, *Nat Rev Genet* 2: 597–606, © 2001 with permission from Macmillan Publishers Ltd.; **2.6,** reprinted from Nicholson KE, et al., 2007, *PLoS One* 2: e274, dewlap images reprinted with permission from David Hillis; **2.7B,C,** reprinted from Ryan MJ, Rand AS, 1995, *Science* 269: 390–392.

Chapter 3

3.1, Redrawn from http://www.philtulga.com/MSSActivities.html, with permission from Phil Tulga; **3.2, top,** redrawn from http://people.virginia.edu/~mk3u/mk_lab/electric_fish_E.htm, with permission from Masashi Kawasaki; **3.2, bottom,** redrawn from http://www.biosci.utexas.edu/neuro/ HaroldZakon/research.html, with permission from Harold Zakon; **3.4,** redrawn from Scholarpedia, Dr. Paul S. Katz, Georgia State University, Atlanta, GA; **3.6,** reprinted from Kyriacou CP, Hall JC, 1980, *Proc Natl Acad Sci* 77: 6729–6733, with permission from C.P. Kyriacou; **3.7,** modified from Witte K, Ryan MJ, *An Behav* 63: 943–949, © 2002 Elsevier.

Chapter 4

4.1, Modified from Ball GF, Balthazart J, *Horm Behav* 53: 307–311, © 2008 with permission from Elsevier; **4.2,** adapted from LaDage LD, et al., *Anim Cogn* 12: 419–426, © 2009 with permission from Springer Science+Business Media; **4.3,** redrawn from Sherry DF, et al., *Brain Behav Evol* 34: 308–317, © 1989 with permission from S. Karger AG, Basel; **4.4, 4.5,** adapted from Ewert J-P, *Neuroethology,* © 1980 with permission from Springer Science+Business Media; **4.6,** based on Knudsen EI, Konishi M, *J Comp Physiol A* 133: 13–21, © 1979 Springer Science+Business Media, and Konishi M, *Cold Spring Harb Symp Quant Biol* 55: 575–584, © 1990 CSHLP; **4.7,** redrawn from Olsen JF, et al., *J Neurosci* 9: 2591–2605, © 1989 with permission from the Society for Neuroscience; **4.8A,B,C,** reprinted from Sichert AB, et al., *Phys Rev Lett* 97: 068105-1–068105-4, © 2006 with permission from the American Physical Society; **4.9A,** source no longer available; **4.9B,** redrawn from Geffeney S, et al., *Science* 297: 1336–1339, © 2002 with permission from AAAS; **4.9C,** redrawn from Geffeney S, et al., *Nature* 434: 759–763, © 2005 with permission from Macmillan Publishers Ltd.; **4.10,** modified from CDC webpage no longer available; **4.11,** redrawn from Krebs JR, *Behavioral ecology: An evolutionary approach* (ed. Krebs J, Davies NB), © 1978 with permission from John Wiley & Sons Ltd.; **4.12, 4.13,** redrawn from Pompilio L, et al., *Science* 311: 1613–1615, © 2006 with permission from AAAS; **4.14,** redrawn from Lemon WC, *Nature* 352: 153–155, © 1991 with permission from Macmillan Publishers Ltd.

Chapter 5

5.1A, Redrawn from Lincoln FC, et al., 1998, *Migration of birds*, Circular 16, U.S. Fish and Wildlife Service, Dept. of Interior, Washington, DC; **5.1B,** reprinted from Egevang C, et al., 2010, *Proc Natl Acad Sci* 107: 2078–2081, with permission from Carsten Egevang; **5.2,** redrawn from Cox GW, *Evolution* 22: 180–192, © 1968 with permission from John Wiley & Sons Inc.; **5.3B,C,** redrawn from Berthold P, et al., *Nature* 360: 668–670, © 1992 with permission from Macmillan Publishers Ltd., and Rolshausen G, et al., *Curr Biol* 19: 2097–2101, © 2009 with permission from Elsevier; **5.4A,** redrawn from Roff DA, *Evolution* 40: 1009–1020, © 1986 with permission from John Wiley & Sons Inc.; **5.4B,** redrawn from Roff DA, Fairbairn DJ, *BioScience* 57: 155–164, © 2007 with permission from The University of California Press; **5.5A,B,C,E,F** (redrawn), **5.5D** (reprinted) from Buhl J, et al., *Science* 312: 1402–1406, © 2006 with permission from AAAS; **5.6,** redrawn from Gwinner E, *IBIS* 138: 47–63, © 1996 with permission from John Wiley & Sons Inc.; **5.7,** redrawn from Chapman J, et al., *Science* 327: 682–685, © 2010 with permission from AAAS; **5.9A,B,** redrawn from Emlen ST, Emlen JT, *Auk* 83: 361–367, © 1966 The American Ornithologists' Union, published by the University of California Press; **5.9C,** redrawn from Emlen ST, *Auk* 84: 309–342, © 1967 The American Ornithologists' Union, published by the University of California Press; **5.10A,B,C,** redrawn from Labhart T, Meyer EP, *Curr Opin Neurobiol* 12: 707–714, © 2002 with permission from Elsevier; **5.10D,** redrawn from Flamarique IN, et al., *J Opt Soc Am A* 15: 349–358, © 1998 The Optical Society of America; **5.11A,** redrawn from Walker MM, et al., *Curr Opin Neurobiol* 12: 735–744, © 2002 with permission from Elsevier; **5.11B,C,** redrawn from Semm P, Beason RC, *Brain Res Bull* 25: 735–740, © 1990 with permission from Elsevier; **5.12A,** redrawn from

Wiltschko W, Wiltschko R, *J Exp Biol* 199: 29−38, © 1996 with permission from The Company of Biologists Ltd.; **5.12B,** redrawn from Lohmann KJ, et al., *J Exp Biol* 210: 3697−3705, © 2007 with permission from The Company of Biologists Ltd.; **5.13A,** redrawn from Lohmann KJ, Lohmann CMF, *J Exp Biol* 199: 73−81, © 1996 with permission from The Company of Biologists Ltd.; **5.13B,** redrawn from Lohmann KJ, et al., *Science* 294: 364−366, © 2001 with permission from AAAS.

Chapter 6

6.4, Modified from Crews D, *Horm Behav* 18: 22−28, © 1984 with permission from Elsevier; **6.6,** redrawn from Hunter ML, Krebs JR, *J Anim Ecol* 48: 759−785, © 1979 with permission from John Wiley & Sons Inc.; **6.7,** redrawn from Seehausen O, et al., *Nature* 455: 620−626, © 2008 with permission from Macmillan Publishers Ltd.

Chapter 7

7.1, Redrawn from Moriarty Lemmon E, *Evolution* 63: 1155−1170, © 2009 with permission from John Wiley & Sons Inc.; **7.2A** (modified), **7.2B,C** (redrawn), from Frishkopf LS, et al., *Proc IEEE* 56: 969−980, © 1968 with permission from IEEE; **7.3A,** redrawn from Huber F, Thorson J, *Sci Am* 253: 60−68, © 1985 with permission from Scientific American; **7.3B,** redrawn from Imaizumi K, Pollack GS, *J Neurosci* 19: 1508−1516, © 1999 with permission from the Society for Neuroscience; **7.3C,** redrawn from Nabatiyan A, et al., *J Neurophysiol* 90: 2484−2493, © 2003 with permission from the American Physiological Society; **7.3D,** redrawn from Huber F, *International Society for Neuroethology Newsletter*, November 2006, 3−6; **7.4A,** redrawn from Brenowitz EA, Beecher MD, *Trends Neurosci* 28: 127−132, © 2005 with permission from Elsevier; **7.4B,** modified from Nowicki S, et al., *Amer Zool* 38: 179−190, © 1998 with permission from Oxford University Press; **7.5,** redrawn from Verzijden MN, ten Cate C, *Biol Lett* 3: 134−136, © 2007 with permission from the Royal Society; **7.6,** modified from Rosenzweig MR, et al., *Biological psychology*, © 1996 with permission from Sinauer Associates; **7.7,** modified from Jones AG, et al., *Proc R Soc Lond B* 267: 677−680, © 2000 with permission from the Royal Society; **7.8,** reprinted with permission from Darryl Gwynne; **7.9,** reprinted from Emlen DJ, *Ann Rev Ecol Evol Syst* 39: 387−413, © 2008 with permission from Annual Reviews, Inc.; **7.11A,** redrawn from Schiestl FP, et al., *Nature* 399: 421−422, © 1999 with permission from Macmillan Publishers Ltd.; **7.11B,** courtesy of Manfred Ayasse; **7.12, left,** redrawn from Petrie M, *Nature* 371: 598−599, © 1994 with permission from Macmillan Publishers Ltd.; **7.12, right,** © Tom Ulrich/Visuals Unlimited; **7.13,** reprinted and redrawn from Kim TW, et al., 2007, *PLoS One* 2: e422, photo courtesy of Taewon Kim; **7.14,** redrawn from Lampert KP, et al., *Curr Biol* 20: 1729−1734, © 2010 Elsevier.

Chapter 8

8.1A, reprinted from Lim MM, et al., *Nature* 429: 754−757, © 2004 with permission from Macmillan Publishers Ltd., courtesy of LJ Young, Emory University; **8.1B,** adapted from Young LJ, et al., *Nature* 400: 766−768, © 1999 with permission from Macmillan Publishers Ltd.; **8.2,** redrawn from Hammock EAD, Young LJ, *Science* 308: 1630−1634, © 2005 with permission from AAAS; **8.3,** based on data from Rosenblatt JS, et al., 1979, *Adv Stud Behav* **10**: 225−311, and Ball GF, 1991, *Acta XX Congressus Internationalis Ornithologici* 984−991; **8.4,** redrawn from Alberts JR, Gubernick DJ, in *Mammalian parenting* (ed. Krasnegor NA, Bridges RS), © 1990 Oxford University Press; **8.5,** redrawn from Liu RC, Schreiner CE, 2007, *PLoS Biol* 5: 1426−1439; **8.6,** redrawn from DuVal EH, *Auk* 124: 1168−1185, © 2007 The American Ornithologists' Union, published by the University of California

Press; **8.7** (redrawn), **8.8A,B** (reprinted), **8.8C** (redrawn) from Goodson JL, et al., *Science* 325: 862–866, © 2009 with permission from AAAS; **8.9,** courtesy of Joan Strassman, Rice University; **8.10,** reprinted from Ryan MJ, et al., *Am Nat* 161: 380–394, © 2003 University of Chicago Press; **8.11A,** redrawn from http://maarec.cas.psu.edu/bkcd/hbbiology/colony_org.html, courtesy of the U.S. Department of Agriculture.

Chapter 9

9.1, Redrawn from Nelson RJ, *An introduction to behavioral endocrinology,* © 1995 with permission from Sinauer Associates; **9.2,** redrawn from http://en.wikipedia.org/wiki/Lek; **9.3,** redrawn from Langen TA, et al., *Behav Ecol Sociobiol* 48: 285–292, © 2000 with permission from Springer Science+Business Media; **9.4,** redrawn from Leiser JK, *Anim Behav* 65: 453–462, © 2003 with permission from Elsevier; **9.5,** redrawn from Oliveira RF, et al., *Proc R Soc B* 265: 1045–1049, © 1998 with permission from the Royal Society; **9.6,** redrawn from Huhman KL, et al., *Horm Behav* 44: 293–299, © 2003 with permission from Elsevier; **9.7,** redrawn from Wommack JC, et al., *Horm Behav* 43: 229–236, © 2003 with permission from Elsevier; **9.8A,** redrawn from Kelley DB, Brenowitz E, *Behavioral endocrinology 2e* (ed. Becker JB, et al.), © 2002 with permission from MIT Press; **9.8B,** modified from Burmeister SS, Wilczynski W, *Horm Behav* 40: 550–558, © 2001 Elsevier; **9.10,** data from Yang E–J, et al., *Ethology* 107: 777–793, © 2001 Wiley–Blackwell; **9.11A** (modified), **9.11B** (data), from Farrell WJ, Wilczynski W, *Anim Behav* 71: 1155–1164, © 2006 Elsevier.

Index

Page references followed by f denote figures; those followed by t denote tables.